O.C. "Russ" Tirella and Gary D. Bates

WIN-WIN NEGOTIATING

A PROFESSIONAL'S PLAYBOOK

Published by the
American Society of Civil Engineers
345 East 47th Street
New York, New York 10017-2398

ABSTRACT

In negotiation, as in most sports, the key to winning is the quality of preparation and practice. A good strategic game plan can overcome tactical errors on the negotiating field much more than in any other sport. WIN-WIN NEGOTIATING: A PROFESSIONAL'S PLAYBOOK, a volume in the ASCE Engineering Management Series, helps the professional to train and prepare the team for the game, how to read and interact with the opposition, and most importantly helps define winning and losing before, during and after the negotiation. The book discusses offensive and defensive strategies, which revolve around what losing and winning mean to the negotiator. The authors make the final point that a winning game of negotiating produces no losers if the game is played right — it's a win-win game for each side. A series of appendices provide various check lists and ideas for improving negotiation skills which have been referred to in th main portion of the book. References are listed in Appendix G.

Library of Congress Cataloging-in-Publication Data

Tirella, O.C.
    Win-win negotiating : a professional's playbook / by O.C. "Russ" Tirella and Gary D. Bates
      p.cm.
    Includes bibliographical references and index.
    ISBN 0-87262-884-1
    1. Negotiation in business. 2. Conflict management. 3. Negotiation. I. Bates, Gary D. II. Title.
HD58.6.T57 1993
658.4—dc20
                               92-46659
                                   CIP

Dedicated to
all the people in the world
who are working to resolve conflicts

Acknowledgments

The authors wish to express their *sincere gratitude* to the many people who made considerable contributions to the overall production of this book:

Pam Giffin, Sharon Butler, and Kelley Carter, who typed and retyped the manuscript and were very patient with our penmanship and many changes along the way.

Zoe Foundotos, our editor, for her encouragement, assistance, and ideas every step of the way, along with Shiela Menaker who guided the manuscript through the production process.

Mel Hensey and Ken Gibble for their comprehensive reviews and suggestions for improvement of the manuscript, as well as John Hribar, Terry McManus, D. W. Ryckman, and Chris Hendrickson for their support.

To the Canada Geese on the lake that inspired us during the final weekend of drafting the original manuscript.

And last, most importantly, to our wives, Marge Tirella and Joyce Bates for all their support on this project and all the other ones that consume us.

Contents

Foreword 1

Part 1
Chapter
1. Holy Cow! It's Going, Going, Gone! 5

Part 2 Getting Ready
2. Headlines for Tomorrow's Sports Page 9
3. Drafting the Right Players 10
4. Who's Shooting to Make the All-Star Game 11
5. Curfew and Other Team Rules 12
6. What's at Stake in the Pot 13
7. How Do You Want the Box Score to Read 15
8. Reviewing the Record Books and Old Press Clips 17
9. Spring Training Takes More Than One Day 19
10. Choosing the Right Playing Field 20
11. Finalizing Strategies for the Game 23

Part 3 Let the Game Begin
12. Useful Tactics 37
13. Techniques for Effective Negotiating Communications 44

Part 4 On Reaching Closure: Tallying the
Final Score and Mounting Your Trophy
14. As the Clock Winds Down 59
15. Post-Game Niceties 62
16. The Post-Game Wrap-Up 64

Part 5 Away Games
17. International Negotiations 67
18. U.S. Government Contracting 70

Part 6—Appendices
Training for Your Next Negotiation

A. Checklist for The Ideal Negotiator 77

B. Ten Primary Reasons for Failure in Negotiation 78

C. Checklist on How to Break Contract Deadlocks 79

D. Checklist for Your Negotiating Effectiveness 80

E. Checklist on How to Improve Listening Skills 81

F. Attitudes Communicated Non-Verbally 82

G. Recommended Reading 85

H. Negotiation Case Study 87

H. Negotiation Case Study 87

Index 89

Foreword

Negotiating is—well—"A Unique Sport." It is also something like a cross between an applied science and an art form.

Since the beginning of time, people, including Adam and Eve, have found answers through negotiation. Negotiating is now a noticeable part of everyday life. From the infant who emits an attention-getting cry, to the grandmother who quiets the crying, negotiation is practiced by all of us daily.

More importantly for the purposes at hand, negotiation is a significant aspect of a design or consulting professional's activities. It is obviously the process that allows an engineer, for example, to enter into a contract to provide design services to his clients. In a more subtle sense, perhaps, negotiation takes place in collecting overdue bills and even in eliciting improved performance from a staff member.

In my experience in the design profession, I have learned—sometimes the hard way—that negotiations must produce a "win–win" result. One of the reasons I like this book is that it emphasizes this point. The outcome of a negotiation, both in fact and in perception, should be a win for both parties. As an illustration, the authors say in the beginning that " . . . the objective of any negotiation is to arrive at a *winning* solution for *both* teams" (italics added).

Win–Win Negotiating: A Professional's Playbook should be of value to the reader for three reasons. First, it is one of the few books on negotiation that has direct relevance to those in the design and consulting professions. Second, it is an easy-to-read, practical book on the theory and process of negotiation. And, finally, it is enjoyable.

McLane Fisher
CH2M HILL
Denver, Colorado

Part 1

Chapter 1

Holy Cow! It's Going, Going, Gone! . . .

"Ball four! Strike three! Touchdown! It's a goal!" When you hear these words you automatically think of a sport. On the other hand, what is the first thing that comes to your mind when you hear "we were selected?" You may think of project, contract, and "backlog." But have you ever thought of the word "negotiate?" If not, we recommend that you do because contract negotiations should be the mound of your "super bowl."

This book is intended to make you aware of the PROCESS. A process that you will continue to learn from, no matter how many negotiations you are involved in and with whom. By reading this book, you will become an expert—a person who knows a tad more of something than the other party.

The dictionary defines negotiation as "to confer, bargain or discuss with a view to reach agreement." A more elaborate definition of negotiation is the art of arriving at a common understanding through bargaining on the essentials of an issue, be it at the office, with family and friends, with clients, with the concessionaire at the ball park, with the person sitting in front of you, or trying to get the autograph of Michael Jordan. Because of the interrelationship of many factors, it is a difficult skill and requires the exercise of judgment, tact, and common sense. The effective negotiator must be a real "shopper," alive to the possibilities of "bargaining" with the seller. Only through an awareness of relative bargaining strength can a negotiator know when to be firm and when to concede.

Negotiation takes place when two parties, each with its own viewpoints and objectives, seek to reach a mutually satisfying agreement on, or settlement of, a matter of common concern. The PROCESS of negotiation involves:

- Each party's presentation of its positions, interests, and needs (PIN); that is, each party's perceptions of what is at stake in the negotiation.
- Each party's analysis and critical *evaluation* of the other's PIN to more fully determine their merits or weaknesses.
- Each party's adjustment of its own PIN to reflect as much of the opposite party's PIN as it considers reasonable and justifiable. These adjustments may result in complete agreement between the two parties; if they do not, a final step consisting of a COMPROMISE by one or both parties may be necessary.

Negotiating is a game—unique in that both sides can and should win. In a normal contest you have a winner and a loser. In a true negotiation, you should have a WIN-WIN situation, which we call a winning game.

On the other hand, negotiating is a SPORT. There are many definitions of sport. According to *Webster's Third New International Dictionary*, one definition of sport is something that is a source of pleasant diversion. Therefore, in order to be successful in negotiating you must have fun doing it. (Here comes the WAVE!)

Since negotiating is an intrinsic part of our every-day life, we must look upon it as a pleasant diversion. Then our "at bats" with our spouse, children; at work with our boss, peers, and associates; with our clients, customers, vendors, regulatory agencies; with airline and hotel representatives; with waiters, taxicab drivers; with friends, neighbors, and relatives; etc., will be more eventful if we learn how to do it in a winning fashion. Playing a winning game of negotiation regularly will bring much more meaning to the game of life.

The purpose of this book is to provide you with some ground rules, ball parks, and hot dogs that will enable you to become a better negotiator. Keep in mind that every negotiating situation is usually a different game. Therefore, do not expect to find a "Louisville Slugger" model in this book. Additionally, you don't have to memorize all the checklists, principles, and batting orders given in this "Professional's Playbook." You will quickly find that basically most of this book is common sense and the rest should come to you naturally (as switch-hitting was to Mickey Mantle). Now, our national anthem: BATTER-UP!

Part 2
Getting Ready

Chapter 2

Headlines for Tomorrow's Sports Page

If you could write the headline for tomorrow's sports page, it would be a statement of what you hope to accomplish in the game today. The same can be said about the negotiating game. You and your team need to know and discuss the results to be achieved and how you will go about achieving them and then write them down. No two games are won using exactly the same strategies  and tactics because every opponent is different, the circumstances are different, and the stakes vary. You must decide HOW MUCH THIS GAME MEANS TO YOU AND YOUR TEAM.* Exhibition games and the first few games of the season allow for experimentation and learning from mistakes but these approaches don't fit in the tournament games. Negotiating a three-year lease on an expensive piece of office equipment is important but not nearly as much as a multi-year contract with a Fortune 500 client to do work on a multiple-site project.

As you and your team plan and develop a statement of what you want to achieve, ALWAYS TALK TO THE TEAM OWNER to see what the game means to the owner and if there are any hidden objectives besides just winning; check to see how the owner defines winning (see Part 3, Chapter 12). Sometimes team owners need to look beyond the pending game for the long-term best interest of the franchise. The owner's position on the winning margin, use of the bench or newly drafted players, how you play the game, and even your reaction to bad calls by the referees or dirty tricks by the opponents are issues that you as the team leader and coach must know. The owner says what is to be accomplished and the coach, with help from the team, develops a game plan to accomplish it. One characteristic all owners want in their coach and entire team is the motivation and optimism that every game can be won. Even when the other team appears more powerful, has a better record, and has more all-stars, you are expected to approach your game plan and execution with only one thought—WE CAN WIN. Remember, both teams should win in a winning game of negotiating.

---

* Not all negotiations are between teams. Some are handled by individuals. When only individuals are involved, they must prepare for and play the game following the same guidelines that are given throughout this book for team play.

Chapter 3

Drafting the Right Players

Signing the right players to fit your team's weaknesses is the key to building a winner. Picking the right members from within your organization for your negotiating team is also critically important. The team leader, knowing what is at stake and what the owner wants to achieve, must pick the talents and skills required to create a strong unit. These player selections must be based on specific needs—NOT ON WHO'S AVAILABLE.

It is important to remember that the size of the team that is allowed to travel to the game site might be much smaller than the team that practices through the week. That does not diminish the importance of having a strong total team to prepare for the game even though some members won't make the traveling squad. Smart coaches will choose wisely from one game to the next so that all players will get some travel and playing experience over the entire season. This practice is critical for the morale and development of the team on a long-term basis.

As you pull your team together and start developing your game plan, remember EVERYONE CAN'T PLAY QUARTERBACK. Every position must be filled. This means not everyone gets to be the spokesperson during the negotiating. Some must do research, some must take notes or keep score, and others need to keep track of the team's equipment and logistical problems. It is up to the coach to make sure that every team member knows the specific role or roles to be played, what is expected of that role, and how that role fits within the overall game plan.

When you approach different parts of your organization to acquire the team members needed for a specific negotiation, remember that the coach needs to WORK OUT THE CONTRACT TERMS WITH THE PLAYER'S AGENT. Most team members will be engaged in other duties when you approach their manager and you must negotiate their release under some agreed-upon terms for the duration (either full or part time) of the negotiating sessions.

Chapter 4

Who's Shooting to Make the All-Star Game

One of the most important duties of the team leader is to examine the expectations of individual team members. EVERY PLAYER WILL HAVE A HIDDEN AGENDA FOR THE GAME. In fact, you as the coach probably do too. It's not enough to be part of a winning team. Most players want to make a contribution significant enough to give their careers a boost. Unless this desire is monitored and controlled by the coach, the team will function as a bunch of individual "hot dogs" and not a cohesive unit. SOME POSITIONS ALWAYS GET MORE PRESS TIME, but every member on the team is necessary to its success. There are a considerable number of games to be played each season and many seasons in a career. Usually, ONE GAME DOES NOT A CAREER MAKE. Sometimes the entire team stays home and only the team leader negotiates one-on-one with the opponent. Success will still belong to the entire team.

Chapter 5

Curfew and Other Team Rules

No team can function effectively without rules to practice and play by. The coach is responsible for establishing GROUND RULES FOR THE TEAM'S CONDUCT. The team leader must act like a boss while leading the team to victory. Input, ideas, and critiques should come from all team members but the final word in practice and during the game belongs to the coach. In the process of developing the game plan and practicing on your own field, it's okay to yell and scream and get excited. This is part of the motivation of getting the team up for the game. However, there are certain rules to be obeyed and these may get more stringent as the BIG GAME approaches. Some suggestions are:

- Each team member should have a defined role and should participate without hogging the show.
- The team should seek to bring out the collective wisdom and talents of the entire team.
- Expect disagreement among team members on some issues but don't take it personally.
- Avoid interruptions or side discussions when others are speaking.
- Make sure that at least one person (perhaps two) is taking notes on the detailed game plan.
- Realize that other team members' perceptions of you are usually helpful, even if you have a hard time accepting them.
- When the team is practicing, the players must speak openly, candidly, and courteously to the coach and fellow team members.
- It is illegal for two teammates to be angry (at each other) at the same time (this carries a fifteen-yard penalty, a technical foul, or possible ejection from the team).

Chapter 6

What's at Stake in the Pot

As we mentioned earlier, one of the most important issues in planning a game of negotiating is to know exactly what is to be achieved. The game may be played between two players and have only one or two issues at stake. Or, it may be played between two sovereign governments or two large corporations and have hundreds of issues  to resolve. It makes no difference. Each team, regardless of size or number of issues, must identify *every* single issue at stake and its priority from its own point of view. Championship teams also try to determine what the other team has at stake on all identifiable issues.

Most teams planning to play a negotiating game concentrate on the financial aspects. Indeed many, if not most issues, have a direct or indirect relationship to money. But certainly not all issues will be tied to personal or corporate profit and loss. THERE MAY BE MORE AT STAKE IN THIS POT THAN JUST YOUR ANTE. Some of the more common stakes besides money that people regularly negotiate for include: ego, pride, tradition, honor, bragging rights, the status quo, dogma, patriotism, life and death, time, quality, turf, family security, etc. Deciding what is at stake besides money is often the most difficult part of planning. Consider, too, that often the informal negotiations that occur every day between two fellow employees or a person and their spouse in a spontaneous situation do not easily allow extensive analysis of what's at stake. You negotiate "on the fly" and realize too late that you're in a game where the stakes could cost you your house. Later on we'll discuss time-outs and other delaying tactics that you (and your team, if appropriate) can use to push away from the game and take another look at the stakes.

Sometimes the stakes change in the middle of the game and your pre-game list of issues needs to be reanalyzed and reworked. Whether you are developing your list before the game starts or reworking it in the middle of the game, get the entire team involved. Don't let anyone on your team play without knowing what the entire team is trying to achieve at any given point. Don't even let your team take the field for a pre-game warm-up without nailing a list of all the chips at stake up on the locker room door.

Chapter 7

How Do You Want the Box Score to Read

It's one thing to have a list of issues that you want to resolve or the results to be achieved, but without specific objectives on each and every issue, you will be trying to shoot an arrow at a bull's eye that's not there. The specific target that you are shooting for must be defined so that there is no misunderstanding of the picture of what winning looks like. It is important to remember not to make the target so small or restrictive that it will be impossible to hit. Be reasonable in the range of acceptability for every target that you set.

Sometimes negotiators try to use the same kind of arrows for all kinds of targets. It won't work. Money issues require arrows that are different from those required by ego issues. That is why it takes so much planning to list all of the issues, set reasonable but specific targets for each, and decide how each should be addressed.

When you start defining ranges of acceptability for each targeted issue, use terms that will relate to that issue in the written agreement at the conclusion of the negotiation. Dollars may be tough to agree on but they are easy to understand. (If the agreement will involve foreign currency, memorize conversion rates thoroughly and learn to think in that currency rather than in dollars.) The targets relating to the scope of work, quality of finished product, work practices, or labor restrictions and time schedules must all be defined in terms understood by both negotiating teams. It is as bad to be shooting at something the other team can't understand or relate to as it is to have no target at all.

If you wish to have a reasonable shot at getting the other team's agreement to the acceptable range of your target on each issue, it is necessary to consider where they might be coming from and whether they will be able to agree within your targeted range of acceptability. This requires your team to develop resistance points on each major issue and guess at the resistance points of the other team.

A resistance point is defined as the point on an issue which you are unwilling to move beyond, without walking away, unless there is a large enough concession on another issue by the other team to justify your going beyond that resistance point. Within the limits of the resistance points is the aspiration levels of both teams. The aspiration level should be a realistic goal or reasonable expectation for an issue targeted for negotiation. The space for negotiation on any issue within the aspiration levels is the winning zone for that issue.

The fewer major league issues there are to be negotiated, the more critical it is for each team to attempt to stay inside the strike zone of concessions when stating early positions on an issue. Taking a position considerably outside the other team's resistance point may provide it with a walk before substantive discussions can be started.

Chapter 8

Reviewing the Record Books and Old Press Clips

Knowledge is power. The more you know about the other team, their true needs, their negotiating limits or points of resistance, their financial or time constraints, etc., the better chance you have of winning your team's objectives and making the other team feel like they won their objectives. Understand, the other team isn't very likely to willingly provide all of the information you require or would like to have. Probably this information will become even more difficult to obtain once the formal negotiations start.

Therefore, it is critical to start well in advance in your research for information that will have a direct bearing on the negotiations. The amount of time spent on reviewing old game records, video tapes, press clips, and other sources of information on both the issues being negotiated, as well as the other team (both corporately and individually), is as important as the outcome of the negotiation. Your team, including the owner as previously mentioned, should know by now how important each negotiating game is to you and you must plan your information research effort accordingly.

As you undertake this research activity, the following questions or statements may be useful reminders of approaches to take or options to consider:

- Have you played against this team recently?
- Were the same players on the other team then?
- Were the stakes in the previous game or games the same as the upcoming game?
- What changes in team members or stakes make this game different for your team?
- If you haven't played against the other team recently or ever, who have they played. Maybe you can borrow some game tapes from them and talk to their coach.
- Talk to people that have played for the other team in the past and been traded, optioned, or released. They may be quite willing to give you useful information.
- Ask the other team directly for all the information you can think of. The worst that can happen, if you ask nicely, is that they will say no. But, they might say yes.

The more information you obtain, the more accurately you can predict the strategies and tactics of the other team. Because of this, be acutely aware that the other team is trying to gather comparable information about your team. Be discreet about what information your organization allows to be disseminated to the media or, upon request, directly to the other team, so that your own needs, strengths, weaknesses, and team goals aren't known by the other team until you decide the best time to give out information as part of your overall strategy.

Chapter 9

Spring Training Takes More Than One Day

Getting ready for any negotiation involves certain steps and activities that must be followed. Just like in the sports world where some position players start training camp early, the team leader must decide when to start the training of some team members and when to train the entire team as a complete unit. Assignments for gathering information or making preliminary logistical arrangements may occur simultaneously with the development of issues to be negotiated and the defining of targets (zone of concessions and the win-win zone for each issue). It is the coach's responsibility to organize the entire process so that all members know their exact assignment. That means not only knowing what they are to do during training camp but also during the game—individually and as a team.

While early training is for team development, understanding individual assignments, and for research and issue listing, at some point the "pads" go on and the team members need to get physical with each other. That means long hard practice sessions where role playing is done and the coach (as well as other team members) critique the performance of their teammates under game conditions. ALL MAJOR NEGOTIATIONS REQUIRE THIS. Not practicing all of your strategies, tactics, and offensive and defensive plays will probably lead to your not playing a winning game. A good rule of thumb for preparing for formal negotiating games is that it takes at least two hours of preparation for every one hour of negotiating with the other team. Of course, the stakes always have to be taken into account when planning and justifying the preparation effort.

Chapter 10

Choosing the Right Playing Field

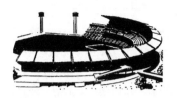 It is always desirable to choose a site that is advantageous to your team. Naturally, the other team has the same idea in mind. Therefore, as a general rule, it is preferable to have a location that eliminates or at least minimizes any major turf advantage for the other team and one that has some advantages for your team. For major league negotiating games, the location of the playing field often becomes a sizable issue of negotiation.

Traditionally, sellers of customized products and services travel to their clients to make presentations and negotiate contract terms. Consumers usually go to retail stores for the negotiation of merchandise they wish to purchase, and job seekers travel to the prospective employer for the interview and job offer. However, in many other circumstances you can strongly influence or perhaps control the setting for the game. If two managers are about to negotiate a resolution to an interdepartmental conflict, a neutral conference room would eliminate any home court advantage. A labor union–management negotiation may choose a local hotel or similar facility for discussions, and two countries may ask a neutral host country to provide facilities for a major negotiation.

The following are some of the issues to consider when deciding on a location:

- What are the travel constraints of both teams getting to that site?
- Is the room (or rooms) out of the main traffic pattern, private (perhaps soundproof), and large enough for the maximum number of participants anticipated to play?
- Are there adjacent rooms for each team to retreat to for private workouts?
- Are there restrooms reasonably close?
- Do all of the rooms have good ventilation, heating or air-conditioning, and exhaust fans, especially if smoking is to be permitted?
- Are there facilities for serving hot and cold beverages and perhaps light meals?
- Are there phones in each room that can have access controlled, especially incoming calls?
- Are there appropriate audio-visual aids permanently or temporarily available in all rooms?

Of course, many of the above are non-issues for small informal negotiations between two individuals or even two organizations where the stakes are small or the negotiation is part of a regularly occurring pattern.

There are advantages to using your own as well as the other team's home field. They are:

Advantages to Playing the Game at Your Field

- It may be easier to get approval from your owner on issues that you didn't anticipate.
- You have the option of handling other urgent/important matters while still participating in the negotiations. (This, however, can be very distracting to your team members as well as to the other team.)
- This will save your team money on travel expenses although there are certainly out-of-pocket costs for hosting the game.
- There is a psychological and emotional boost in having the other team come to you and in playing in front of your fans.
- If the away site involves a distant field, then jet lag, strange beds, and new food can greatly affect your team.

Advantages to Playing the Game Away From Home

- You can devote your time and concentration to the negotiation without interruption.
- As a delay strategy, you can use the fact that certain critical information "is back at the office."
- You can imply your intention to go over your opponent's head to access the team's owner.
- The burden of logistical preparations and interruptions is on the other team.
- It will be easier to obtain approval from the other owner for issues not anticipated.

Consider all of these factors, as well as those that are specific to your circumstances and the size of stakes being negotiated, before readily agreeing to any location.

Seating for negotiating sessions can really impact the atmosphere and be a major contribution or barrier to a winning game. Certain arrangements suggest subordination and can be intimidating to one person or an entire team. In an office setting, the person sitting in the huge executive chair behind the desk often comes across as having a perceived power edge. Some executives, trying to exert their authority or effect a win-lose outcome, have been known to intentionally arrange their office so that their guests are subtly put down by the location or level of the guest seating.

In a neutral conference room setting, experienced tough-minded negotiators often will make it a point to secure the "power seat," which is at the head of the table, with their back to the window, facing the door. The purpose of this is to get a psychological edge in controlling the negotiating process. None of these little psychological power plays will be effective if you don't let them. You may even want to suggest that the positions be rotated as the negotiations proceed.

It may be difficult to determine any one best time to play the negotiating game. For sure, some teams and players do better in day games, while others like to play under the lights. You and your team need to decide when you are at your physical and psychological peak; your adversary will be making the same decision. Usually, we must compromise on these criteria and work within other constraints such as the availability of the team players and deadlines for completing the game. There are some occasions when negotiating should be avoided at all costs. Do not negotiate

- when you are caught totally unprepared to negotiate (delay, postpone, or cancel the game);
- when the other team appears preoccupied with other matters and isn't prepared to give your game its undivided attention;
- when either or both teams are angry or emotionally upset;
- when one or both teams are fatigued—whatever the reason.

All of the foregoing issues on location, room accommodations, and timing will not be factors in every game of negotiating but at least consider them before dismissing any. The game will always be a better one if the choice is the best field for both teams. Only then will this become "a field of dreams."

Chapter 11

Finalizing Strategies for the Game

Two Strategic Decisions

Every game of negotiating requires a unique strategy. As the final step in getting ready, you must use all of the information that you have collected and the understanding of your specific objectives to strategically finalize your approach. Good tactics score points and keep the other team from scoring or scoring only on agreeable terms. Good strategies overcome tactical mistakes to produce winning games and winning seasons. The two most strategic decisions to make before the game starts are:

1. **What is the true value to my team of a good ongoing relationship with the other team?**
2. **What will really constitute a feeling of winning in both teams?**

RELATIONSHIPS count in 99% of all meaningful negotiations. The tricky question is, how much do they count. The reason that teams play the negotiating sport is that they each believe they have something the other needs or desires. Unless you are haggling (this is not true negotiating) with a merchant in a far-away land over a take-it-with-you-right-on-the-spot souvenir, you probably should be very concerned about how the negotiation will affect your relationship.

Buyers as well as sellers of services, products, and ideas need to consider the other team as a customer because customers only deal with people with whom they wish to have a continuing relationship. The exceptions are teams that are desperate, lack self-confidence, and don't have much of value to offer the marketplace. There are very few situations where a good relationship with the other team is unimportant after the negotiating game is over.

You will find intimidation, deceit, manipulation, emotional outbursts, outrageous positions on issues, threats, fear, guilt, abusive language, and total ruthlessness to be very effective tactics for causing anger, resentment, and retaliatory measures from the other team. It usually doesn't require using all of them in one game. Sometimes using just one or two of these tools is enough to win that particular game but don't expect to go out and celebrate afterwards. If the other team has any self-respect at all, that's the last time they'll play the negotiating sport with you and will probably try to "get even" any way they can.

The approach to use in keeping a good relationship or even strengthening one with the other team requires an entirely different attitude before, during, and after the negotiating game. The key to developing this approach is based on a very simple assumption: **One team does not need to dominate the other; in order for one team to win, the other does not have to lose.** Simply put, both teams can and should be winners. Once developed, this approach will allow you to believe that somehow, someway, an agreement can be reached that satisfies the needs of both teams. If the relationship is important, the emphasis is focused on defeating one or more mutual problems and not defeating the other team. We will discuss further the key elements of how to use this new attitude to keep and strengthen the relationship with the other team.

Let us now look at the second strategic question: What really constitutes a feeling of winning in both teams? THE SPORT OF NEGOTIATING SHOULD ALWAYS BE PLAYED TO WIN. However, what constitutes a win is a matter of perception in the minds of the teams and players engaged in the game. Perception is based totally on feelings and feelings are based on all of the past experiences of every player. Think of perception at its best as a person's feelings about confirmed facts. At their worst, perceptions are a person's feelings about untested assumptions. These feelings in either extreme will be the result of how each individual has mentally processed every experience he or she has had in life.

That is why it has been said many times that we don't live in a world of facts—we live in a world of perceptions. The sport of negotiating is certainly founded on this principle. This may be so important to the game that it's worth repeating; **"What constitutes a win or a loss by either team is not based on the facts of the outcome as much as it is based on the individual and collective perceptions of each team regarding the outcome."**

Seldom, if ever, is the final score (negotiation settlement) or any of the points scored during the game, so factual in reality that the true value is obvious to all. Rather, it is almost always both teams' perceptions of that score which determines whether the outcome is a win for both sides. Furthermore, perceptions always come into the play before and during the game, not just at the conclusion.

Regardless of the facts and subjective information that we have collected and put in our game plan, in the final analysis it will be the other team's perception of our game plan and how we present it that will decide whether we reduce or increase our differences at the negotiating table. We must, therefore, always try to see our complete game plan, including all information and how we will present it, through the eyes (or perceptions) of the other team and not just our own.

An item worth constant consideration is checking and calibrating the assumptions that your team has made about the other team and attempting to calibrate what assumptions the opponent has made or is making about you.

One of the reasons negotiating is such a unique sport is that it is the only one where how you play and conduct yourself during the game has as much to do with the feelings of winning or losing as the final score. Scoring on every objective that you set forth in your game plan is important. Assuming you "score" within the acceptable target range for an issue, you give yourself a win for that period of the game. But what if the "score" is outside of the acceptable target range of the other team for that issue. They will feel like they have lost—unless you have made them feel like a winner during this negotiating period. Remember, sometimes the team that goes into the locker room at half-time with too big a lead gets complacent and the other team, with its pride at stake, turns the game around in the second half. Therefore, the most important strategy in this game is, no matter what the score is, whether it's early in the game or at its conclusion, NEVER MAKE THE OTHER TEAM FEEL LIKE IT IS LOSING OR HAS LOST THE GAME.

Offensive Plays

Now that we know the two most important strategic decisions that must be made before we finalize the game plan, the OFFENSIVE PLAYBOOK must be discussed and how it fits into our game plan. The playbook for offensive maneuvers seeks to overcome the obstacles to scoring while at the same time preventing the other team from believing that they are losing.

One of the most common obstacles to a "winning" game is the matter of pre-stated positions. That is why it is dangerous to read too many of your own press clippings. After the media (or your boss) has told the world how great your team is and what prolific scorers or defenders you are, you have a position to uphold. Thus, when opposing teams have publicly (or privately with leaks) stated their positions in advance, they often become locked into those positions to save face. It is extremely difficult to move either team away from its position or claims because ego has become involved. No one likes to lose face.

For this reason, from the planning phase and into the actual game itself, it becomes critically important for each team to focus on the underlying interests rather than the boastful bluffing found in the pre-stated positions. This is very difficult to do and requires the skilled negotiator to separate the projected personalities of those involved from the true problems or interests. **The key is to be soft on the people involved but tough on the underlying problems or issues.**

Four important steps to developing good offensive plays to accomplish this are:
1. Defining the Problem Correctly
2. Determining the True Needs Involved
3. Building Trust and Commitment
4. Creating a Consensus Approach to Problem Solving

Defining the Problem Correctly

Defining the problem or conflict is the very first step in resolving the issue through negotiations. Frequently, too little time is devoted to this step and what is defined is not the problem but rather a symptom. For example, an engineer may define decreased productivity or extreme resistance to a procedural change on the part of his employees as a problem when, in fact, these may be only symptomatic of deeper problems. The following will help you identify true problems:

- Get opinions as well as facts—often opinions put facts into perspective.
- Probe the language used to describe a problem or conflict—be sure both parties are talking the same language. Remember, many words have different meanings to different people.
- Some problems have more than one cause—anticipate the possibility of multiple causes and even multiple symptoms.
- When the apparent problem is defined, probe a bit deeper to be sure it's the true problem. (It is even best to always be a little suspicious of any problem as initially presented.)

Few things are as futile as attempting to successfully negotiate the wrong problem. This is about as futile as trying to hit a golf ball with a tennis racquet.

Determining the True Needs Involved

All negotiations really take place between individuals, even when sizable teams are involved. These individual players either represent themselves alone, or themselves and the team they play for. In either case, the challenge is to find out what motivates the participants into taking the positions that they do. Skilled negotiators spend quite a bit of time questioning, searching, trying to find out what their adversary's true needs are. Sometimes they are not the needs being expressed by the adversary but rather hidden, even intentionally disguised, needs.

To better prepare the negotiator to recognize true needs, let's briefly look at Maslow's hierarchy of needs. Abraham Maslow was a clinical psychologist who categorized all human needs into five groups which he described from the lowest, most basic need to the highest level of aspiration as follows:

BASIC SURVIVAL: the physiological need for food, water, shelter, air, clothing, and basic physical comforts

SAFETY: the need for mental and physical security, including health care and work stability

A SENSE OF BELONGING: the need for love and affection of certain people, the caring and respect of peer groups, and being on someone's team

STATUS: the need to satisfy one's ego through esteem, recognition, and prestige of accomplishments and the praise of others

SELF-ACTUALIZATION: The highest need of all is to feel a total sense of inherent well-being, to personally grow and develop to one's fullest potential. At this level there is heavy emphasis on individuals giving back to society their time and money toward helping others.

Maslow suggested that these needs were important to us in the order given above. The basic need of survival is most important, but once that need is reasonably well satisfied, we strive to satisfy the next highest level of need: security. Once that level is reasonably well satisfied, we strive to satisfy the next highest level: belonging. The pattern continues to the highest level, self-actualization, a level only a small percentage of people reach.

There is another approach to understanding the motivations and needs of those with whom we are negotiating (at times also very helpful in understanding our own motivations as well as those of our teammates). That is to learn to read their MINDS. Each of these letters represents a primary force driving an individual toward a goal:

M: *material* wealth, money, tangible goods

I: *indicators* of status, image, prestige, and saving face

N: *necessities* of life: security (physical, mental, financial), comforts, and peace of mind

D: *desire* for power, to have control over others, to freely exercise will, to have the last word

S: *specifics*, the need for all the facts, details (even minutiae), specifications, and certainty. (This is generally a strong need for people with a background in the physical sciences.)

This approach is slightly different from that of Maslow in that it does not have the sequential relationship of moving from one level of need to another. Most people will be driven by one primary MINDS motivator during their adult life. However, when the stakes are high enough in a negotiating game, some people have been known to change their primary driver. Sometimes the change is just for the duration of the game and sometimes it's permanent.

The Maslow approach or the MINDS reading should allow you to better understand your opponent's true needs and motivators. The reason for the negotiation in the first place is to try to influence the behavior of those from whom we want things, be it money, a contract, recognition, love, job security, or whatever. Several different factors can often satisfy the same need. It is critical to identify and even list the needs that will come into play during the negotiation as early as possible since the entire objective of a winning game of negotiation is to arrive at a solution where the needs of both parties will be satisfied.

Building Trust and Commitment

As we discussed earlier in this chapter, we are making the assumption that the majority of individuals or groups with whom we negotiate are likely to be those with whom we desire to maintain a good continuing or repeating relationship. This is not always true but it is a critical assumption for our offensive playbook in a winning game of negotiating.

It follows that these relationships will be a lot healthier and mutually more productive if they are based on trust and commitment rather than on resentment, deception, and antagonism. It is, therefore, strongly suggested that a continuing trusting relationship be established well before any negotiation takes place rather than attempting to accomplish this during the negotiation. The initiative is yours. The more you trust others and communicate that trust, the greater the likelihood of them responding in kind. This in turn will increase respect and commitment between the negotiating parties.

There are three key factors to building mutual trust:

1. The perception by others that you have the knowledge, experience, and authority to make your statements and positions valid and credible.
2. The perception by others that you are willing to tell the truth (even if you don't tell everything you know).
3. The perception by others that you are interested in them as human beings, their needs, concerns, and interests.

Creating a Consensus Approach to Problem Solving

A revolution is sweeping many organizations in this country. This revolution is in allowing and empowering workers to participate more fully in the work-related decisions that affect them and their organizations.

This movement was stimulated in part by the success of similar philosophies practiced overseas through the work of Dr. W. Edward Deming and others. Programs such as the General Motors *Quality of Work Life*, Motorola's *Participative Management*, Chrysler's *Worker Participation Program*, Champion

Paper's *Worker Involvement*, and Northrup's *Quality Circles* all rely to a large extent on greater worker participation in problem solving and decision making. The success stories evolving from some programs such as these are indeed impressive.

A common denominator to the success of such programs is the ability to bring together individuals from various disciplines, skill levels, and positions and hammer out consensus solutions to work-related problems. How does one achieve group consensus?

It is important to understand that all parties involved in group consensus must actively discuss the issues surrounding the decision to be made. The group thus pools the knowledge and experience of all its members. Out of this collective pool of knowledge can then come a more enlightened decision. The decision must be supported by each member. This approach allows the ideas and the feelings of all the members to be integrated into a group decision rather than producing a we vs. they standoff.

Decision by consensus is usually difficult to attain and will usually consume more time than other methods. As the energies of the group (both negotiating teams) become focused on the problems at hand (rather than individual points of view), the quality of the decision as well as the level of acceptance is elevated. This conclusion is supported by research as well as practical experience. Consensus problem solving is essential to achieving a winning game of negotiating. The steps in this process are:

- GENERATE A NUMBER OF POTENTIAL SOLUTIONS. It is often difficult to come up with a good solution quickly, but those initial solutions may pave the way for better ones. Be sure to get your adversary's possibilities early on. It's not necessary to stake out your position first. Discourage evaluation until a number of solutions have been proposed. Brainstorming may help at this point.

- EVALUATE THE ALTERNATE SOLUTIONS. Both parties should do a lot of honest reflecting at this point. Test all possibilities critically. Any reasons why they won't work? Are any too hard to implement? Do any not satisfy the needs of both parties? Try to eliminate as many points as possible during this stage.

- MAKE A DECISION. A mutual commitment to one solution is essential. Don't try to unduly persuade or force a solution on your adversary. Usually at this point one solution stands out as superior.

- IMPLEMENT THE SOLUTION. Often the ball gets dropped at this point. It is essential to decide who does what by when.

- CONTINUE TO MONITOR THE SOLUTION. Often a solution may not turn out to be ideal or 100% effective. Unless there are contractual restrictions, the players can continue to fine-tune the solution closer to the ideal.

Even though many of the concepts brought up in this chapter won't be played out until after the game has started, if you don't strategically plan how and when to use them and put them in your offensive playbook before the game starts, it may be too late once the action begins.

Defensive Plays

Your DEFENSIVE PLAYBOOK is just as important as your offensive playbook from the strategic point of view. There are specific tactics to slow down or stop a particular scoring drive. We will discuss those in the next chapter under TACTICS.

Here we are talking about those defensive strategies that will prevent you from losing the entire game. As in most other sports, the best defense is a great offense. It is imperative to remember that your offensive strategies are set up to ensure that your team scores in the target range of each objective. Likewise, the defensive strategies and the corresponding defensive tactics are necessary to make sure the other team doesn't score outside of your target range for each objective. **Simply put, offensive strategies define a winning solution on every issue and how to reach it, and defensive strategies define losing on every issue and how to avoid it.**

This is where definitions get tricky. Most people think that losing is the opposite of winning. In most sports that's true. But not in negotiating. Losing is just like winning in that it is only losing when it is perceived to be losing by one or both teams. The winning and losing positions on any issue are totally related to how those positions are viewed. Therefore, when you are making a concession on some point and feel that you are allowing the other team to score, it doesn't mean they feel that they are winning. The degree to which you were willing to move off your original position to "help them" may not have been enough movement to get them inside their acceptable target range for that issue; therefore, they still feel like they're losing on that score.

You may in fact have to move to the limits of your acceptable target range on an issue or even go outside the range to get them inside their target range. So one of the key elements of defensive strategies is to examine the issues just like you did for your offensive strategies and see where you *might* be able to expand your range of defined winning without *feeling* like you are losing. This should be done in advance on all major issues and many minor ones so that you don't have to call time-out to do this during the game.

This leads to the conclusion that the best and only true defensive strategy is to plan before the game starts at what position, once the other team proves immovable on any issue, is your team prepared to move outside your target range without feeling like losers or stopping the game.

Because most negotiators do not plan their defensive strategy at all (or at least not as well as their offensive strategies) they become victims of the other team's powerful offense and either *feel* like losers after the game or *are indeed* losers but couldn't walk away from the game. Negotiating is the one and only sport that you do not lose by forfeiting when you walk off the field—DEPENDING on how you define losing.

Using Your Power

The last stage of getting ready is to know and PRACTICE USING YOUR TEAM'S STRENGTHS AND WEAKNESSES (POWER BASES). These may have come out during game preparation and may already be factored into your offensive and defensive strategies. This is the last chance to see if you've covered all of your bases—or are still standing in the batter's box.

As we've said before, the objective of any negotiation is to arrive at a winning solution for both teams. The chances of this happening are much greater if you are able to control the negotiation—not dominate but control. Control comes from power. Three particular sources of power that are important in negotiating are time, information, and leverage. Always use a portion of your practice time to evaluate your strengths and weaknesses, as well as your opponent's, in these three vital areas.

Time

The amount of time available (or perceived to be available) for each party to conclude the negotiation frequently influences in some way the concessions each is willing to make.

For example, on the last day of a special one-month sales blitz, a salesperson may still be a few units short of the assigned quota. Since sizable bonuses await those who exceed their respective quotas the salesperson may feel increasingly obligated to make some extra concessions as the deadline approaches. This is true not only of buy-sell negotiations but others as well.

Four key elements to consider in using time as a power base are:

1. It is to your advantage to know the other person's or team's deadlines whenever possible.
2. It is not to your advantage for your opponent to know your personal or team's deadlines.
3. Often major concessions and settlements are not agreed to until the deadline actually arrives: 80% of the concessions may come in the last 20% of the negotiations. To maintain a calm and poised demeanor is always helpful during negotiations, but especially so as the deadline approaches.
4. Whenever possible, you should set the deadline for concluding the game.

Information

As previously mentioned, KNOWLEDGE IS POWER. The more you know about the other party, their true needs, their acceptable target ranges, their time constraints, etc., the greater are your chances for winning. It isn't likely, however, that your opponent will willingly give you all of this information. The probability of it being available to you once the negotiating game starts is greatly reduced. This suggests starting well ahead of time in getting the information that will have a direct bearing on the negotiations. The amount of time spent in this endeavor should be roughly in proportion to the importance of the negotiation and its desired outcome.

Remind yourself and your entire team that it is almost always advantageous to receive information before the game and seldom to your advantage to give information in advance unless you have carefully leaked the information that you want known about you to the press (or the other team directly).

Leverage

In the organizational world, leverage has come to mean receiving multiple benefits from the assets one has. So it is in negotiations as well. The exertion of a relatively small amount of power at critical times during a negotiation can result in a payoff many times  greater than the effort expended. Five of the most important kinds of leverage used in negotiating games are:

1. *The Leverage of Expertise:* If one team in the negotiations is able to come into the meeting with pre-established credentials in one or more of the issues being negotiated, or if that team quickly establishes those credentials in the early part of the game, the level of respect rises with a commensurate shift in power. This suggests, therefore, that if that expertise doesn't exist, some time ought to be spent in boning up on that subject. If this is not possible

because of time or other constraints, consider bringing along your own expert or experts on various issues (refer to Chapter 3). If your players have been in enough All-Star games, their credibility will be easily perceived and the other team's respect instantly forthcoming.

2. *The Leverage of Investment:* There is a strong human tendency against wanting to lose any or all of an investment. This seems to be true whether the investment is time, money, relationships, or whatever. The more time or money already committed, the more likely we are to make a compromise rather than lose the investment.

    Skilled negotiators are keenly aware of this tendency. They know that if opponents have spent hours negotiating several points of an issue they are more likely to concede on the last point rather than risk a deadlock and perhaps lose all of their invested time. This is why it is usually wise to save the more difficult issues until the last few plays of the game.

3. *The Leverage of Patience and Persistence:* Many who come out on the losing side in a negotiation do so because they aren't patient or persistent enough. They are looking for the quick fix and the patience and persistence of the opponent invariably win out. Patience and persistence are among the most powerful tools a negotiator has to work with. Few areas illustrate the principle of leverage as dramatically as the situation where just a little more patience or a little more persistence can and does bring about much large payoffs.

    Remember that the game is not over until it's over, unless you (or the other team) say it's over.

4. *The Leverage of Attitude:* Attitude is probably the most important leverage to use in any negotiating game. Frequently those who enter a negotiation with a negative attitude find that it will negatively influence one's perception of the issues, the level of cooperation between participants, the rate of progress toward a conclusion and, ultimately, the quality of the outcome itself.

    On the other hand, those who enter the negotiations with a healthy positive attitude demonstrated by an open mind, self-confidence, poise, understanding, and enthusiasm will discover rewards far beyond the effort expended.

5. *The Leverage of Legitimacy:* Legitimacy is an extremely powerful leverage in negotiation. It occurs when one party is able to successfully align its proposal or position with either an accepted standard or established procedure or by what has in the past been customary or traditional. When one team says "that's the way it has always been done around here" or "those are the

only contract terms we're allowed to use," the other team will instantly feel as if it has lost unless it has considered how to deal with this in its defensive strategies. The use of legitimacy usually leaves no room for creative new approaches, especially if the team using this leverage doesn't care about the other team feeling like losers or even who the other team is.

The following are a few additional items to consider in analyzing your team's overall strengths and weaknesses before going into the game:

- Your resources to satisfy the other team's needs are more available or located closer than those of your competitors.
- You happen to know that your team has been recommended to the owners of the other team by a close personal friend of their own.
- Your team knows all or most of the other team's members because you've played in many games together.
- Your primary competition (usually ready to play if you forfeit the game) has recently had major financial difficulties and is not in any position to compete should the other team use that threat.

There can be many more factors in any given negotiating game that can be used as power bases, depending on the circumstances. You and your team must be totally honest in the evaluation of your strengths and weaknesses. What you perceive as a power base can turn out to be a strike out depending on how it is used or perceived by the other team.

Part 3
Let the Game Begin

Chapter 12

Useful Tactics

As we remarked earlier, sometimes the entire team doesn't travel to away games, but the STARTING TEAM and key reserves must always go. The size of the squad will depend on how important the game is. The starting team should be announced no later than the last practice and even before, if possible, so that everyone can practice playing their position. Any last minute scratches from the starting team due to injury, illness, or (the unthinkable) other priorities must be replaced immediately with another practiced team member, even one who is not as skilled as the sidelined player. At this point, knowing the game plan is far more important than slightly better skills.

It is fair to say "DON'T LET ANYONE ON THE PLAYING FIELD WHO DOESN'T KNOW THE GAME PLAN." If there is a last minute substitution, the coach must be certain that player is provided with a written copy of the entire game plan for a crash study before the game starts. This is why it is so important to document all issues, acceptable target ranges, offensive and defensive plays, research on the other team, and all other decisions made during practice sessions. To say it even more strongly: In major negotiating games, it is better to be a player or two short than have a full team that doesn't know the game plan.

The game itself will normally go through four successive stages: (1) Introductions and Rapport Building; (2) Agenda and Logistical Considerations; (3) Discussion and Issue Bargaining; and (4) the Wrap-Up.

As in other sports, it is very appropriate for all of the players to be introduced with a few brief statements about the background and position of each player. This can be done by the coach or by the players themselves. It can be very effective for each team to distribute a written handout on its players at the beginning of the game or, in some cases, even to the other team in advance of the game. This is especially desirable when you are using experts and their credentials are critical to the game.

Do not attempt to rush through this introduction/rapport building phase. Getting to know each other well and building relationships are very important to a winning game of negotiating. It is conceivable that lengthy negotiations on a neutral field may warrant social or recreational activities to be attended by both teams.

A negotiation game is like any meeting. There should be an agenda. It works best when the two team leaders have communicated in advance regarding the agenda as well as logistical needs. This probably will not be a major consideration with one-person teams but becomes a critical factor with large teams, when there are many complicated issues to resolve, and when playing on a field that is home to neither team.

Any prior work on the agenda and logistical accommodations can help create the right environment for a winning game of negotiating. The converse is also true. Whether or not pre-work has been done in these areas, before any discussion of substantive issues begins, an agenda with built-in flexibility should be finalized and any questions on facilities, support services, travel issues, or whatever should be addressed. To the extent possible and practical, all of this kind of information should be in writing and a copy prepared for all players.

As the opening ceremonies conclude and the game finally starts, the only certainty about negotiating is its uncertainty and unpredictability. If all of the other team's players were rational, logical, open-minded, and reasonable, then they would act and react in fairly predictable patterns to our offensive and defensive plays. However, that description does not fit many people when major stakes or conflicts are on the table. Thus, we need to be very flexible in the use of our tactical moves. It is yet another reason to have planned solid offensive and defensive strategies. Unless your acceptable target ranges are too narrow, your perceptions of winning and losing are distorted, or your view of the on-going relationship with the other team is misguided, your strategies should not require drastic changes during the game.

You should count on your tactics needing to change and be prepared to do so. This means that the part of the game plan dealing with individual plays should have room to add new plays or tactical moves as the game progresses. It would be naive to lay out exactly what to say and do on each issue and not be prepared to deviate from that script. There is never one approach on an issue, even if you've played the opposing team before, that will always work. Conditions and people change. So, too, must tactics change. There are some principles of tactics that have been found to have a fairly high success rate. Let us look at a few of them.

For the opening moves and first few plays consider:

A. *Credibility and Believability:* The early stages of most negotiations are marked by a period of testing the waters (like boxers do in the first round).

Even informal discussions not directly related to the issues may be an attempt to establish early dominance. It is important from the beginning to establish and maintain one's own credibility and believability as a basis of mutual trust. Credibility refers to the truthfulness of individuals whose statements can be checked or later verified for accuracy. Believability is an assessment of how true a statement is when it can't be easily checked. Common sense suggests and studies have shown that those individuals who rate high in credibility are much more likely to get their way than those with a low rating.

B. *Identifying and Clarifying the Issues:* Both teams must be sure that they are discussing the same issues. There is often a tendency not to allow enough time for, or even skip, this important step entirely. For the more important negotiations, it is helpful to actually list the issues on a blackboard or flip chart so that both sides are viewing the same thing. This is a good approach even when the issues may be listed on the agenda. When written where all can see, the issues can be fully expanded rather than written in abbreviated form as is often done on agendas. This reduces the possibility of misunderstanding considerably and allows the negotiation to get started on a note of agreement. For buy-sell negotiations, the use of inexpensive visual aids is useful not only to clarify the presentations but as a psychological aid as well.

C. *Initial Concessions:* Experience has shown that in a large percentage of negotiating games both sides are reluctant to make the first concession lest this be perceived as a sign of weakness by the opponent. Actually there is an advantage in making the initial concession providing it's a relatively small one. By taking the initiative, control can shift in your direction. Further, it establishes you as a positive, cooperative negotiator, ready and willing to get on with the proceedings. It also puts a degree of pressure on the adversary to reciprocate in some manner.

This brings us to a useful negotiating tactic called GRIT. GRIT stands for Gradual Reduction in Tension. In using the GRIT approach, one side makes a small unilateral concession. This shifts the pressure to the other side to make a similar concession, and this procedure goes back and forth, gradually reducing the tension between the parties. GRIT seems to work best on the easier issues and in doing so quickly defines and isolates the tougher issues.

d. *Defer Tough Issues:* Several reasons to defer the issues, until the latter part of the game, that are expected to create the biggest disagreements are:

- You will have had an opportunity to demonstrate a positive cooperative attitude in the negotiations before the tough issues are considered. This feeling between both teams is essential for winning negotiations.

- If the winning game approach is firmly established by both teams during the resolution of the minor issues, the chances will be good that that approach will be used for the tougher issues as well.

- You will have an opportunity in the early stages of the game to further define the adversary's needs, tactics, strengths, and weaknesses through the use of questions and body language observation. This additional information will become even more valuable as the negotiations become tougher. The early plays should always be used to test the assumptions that you have made about the other team in preparing for the game.

- The pressure to reach a settlement will build as the end of the game approaches. Experience again has shown that most of the truly meaningful concessions are made near or even after the final whistle. Those who have discipline and patience can profit from this knowledge.

E. *Using the Concession Stand:* In a win–win negotiation, concessions by both sides are inevitable. Successful winning negotiations should meet the needs of both teams. This doesn't necessarily mean the opponent must be on the receiving end of the majority of the concessions or even win the most important points discussed. Many times the way that the concession is made is even more important than the true value of the concession itself. The following suggestions can be used to significantly increase the satisfaction level of any concession granted to the other team.

- **Do not agree to any concession too quickly or easily.** Even if it's a point that has little value to you, your opponent will generally respect your considered response and will frequently place a higher value on the concession as a result.

- **Ask for something in return whenever a concession is made.** This approach tends to make the other team put a much higher value on what you give because now a price has to be paid for it.

- **Aim high but not too high to totally scare off the adversary.** This gives you a wider negotiating range in which to maneuver. Often the opponent perceives this as receiving a greater concession than that actually made.

- **At the resolution of each issue and at the end of the game, compliment your adversaries on their skill and ability.** Reinforce their belief that they got a "good deal." Making them feel good at the end paves the way for future successful negotiations with the same team.

The golden rule in using the Concession Stand is: ASK FOR what has high value to you at low cost to the opponent; GIVE UP what has high value to the opponent at a low cost to you.

A tactic that hopefully it won't be necessary to use is the ULTIMATUM. An ultimatum carries with it the inherent danger of being threatening to the other team and can always be considered a high-risk tactic. Should the opponent reject the ultimatum, you are faced with the often unpleasant task of carrying out the conditions of the ultimatum or threat in order to remain credible. Should the adversary accept the ultimatum, it often is interpreted as losing, thereby compromising a winning game. The adversary may even feel bitter and look for ways to retaliate or get even in the future.

Once in a while, however, the ultimatum may represent the last and only alternative to resolving a conflict. Even though it's difficult, it is possible to use the ultimatum and avoid the difficulties described above. Consider the following steps in using the ultimatum.

- Make sure the opponent has no other alternatives or way out. A take it or leave it price to a buyer will have little effect if the buyer has a secret second source at a comparable price; the ultimatum will predictably fall through and so will the negotiations.
- Soften the wording of the ultimatum as much as possible. "I've got you now" attitudes or harsh "take it or leave it" statements tend to generate defensiveness and negative emotions. Make it as easy as possible for the other person to accept your ultimatum without losing face.
- Often, the ultimatum can be presented along with a second even more unattractive alternative (the "lesser of two evils"), giving the opponent at least some minimal satisfaction in accepting. By doing this you are giving the adversary a choice and thus, the opportunity to save face.

Another tactic that involves both offense and defense is DETECTING and RESPONDING TO RESISTANCE. You can be quite sure that a resistance move is about to be encountered when you hear or see the other team behaving as follows:

- Their body language shows total inattention to what you're saying.
- The opponent constantly interrupts with trivial, irrelevant questions or statements.
- It is almost impossible to get a response to your questions.

- The opponents repeatedly make petty criticisms of you, your company or products or services.
- The opponent cuts short your presentation of an idea or solution.

The three words that best represent all forms of resistance are: STALLS, ALIBIS, and OBJECTIONS (SAO). These are widely used defensive tactics and the closer you get to resolving an issue, especially straight buy-sell negotiations, the greater the tendency for the other team to use an SAO. If we lose our composure and forget our game plans when the other team uses one or more of these tactics, we become an ineffective negotiator and stand a good chance of not winning the game. The anxiety of an opponent resorting to SAO is a natural defense mechanism against making a decision. We must move offensively to relieve their anxiety and then their resistance will decrease.

Following are important steps to responding to SAO's or any other kind of resistance:

- Don't attack the opponent personally or their resistance statement.
- Consider the possibility that you or your teammates may have created the basis for the resistance, probably unknowingly.
- Begin with a non-threatening approach such as acknowledgement of the problem or concern; indicating that you understand an SAO, doesn't mean you agree with it.
- Through questions, determine if the resistance is personal (based on emotions) or objective (based on facts). The latter is much easier to handle.
- Always keep the conversation focused on the factual aspects of the objection rather than on the opponent's subjectivity, personality, unreasonableness, etc. Criticizing the other team seldom overcomes resistance and generally solidifies it.
- Don't use "yes, but" statements.
- Look for hidden objections; these are much more insidious than stated objections.
- If nothing else succeeds, try to brainstorm creative alternatives that go over, around, or under the objection, or at least minimize it.

As the game progresses and issues are resolved, ACCURATE SCOREKEEPING becomes critical. DO NOT WAIT UNTIL THE END OF THE GAME (unless it is a *very* short one) TO TALLY THE SCORE. After each issue is resolved, mark the score in two places. First, record it on the blackboard or flip chart where the issue was written. Make sure that *all* terms and details relative to that issue are publicly documented. This will save much disputing and consternation later on when one or both teams may forget the details of the scoring play. Second, if there are two or more players on your team, assign one of them as scorekeeper. The

scorekeeper will not only write down everything agreed to publicly, but will document answers to questions from both teams, make comments on each team's perceptions of reactions of the other team to tactics and questions, and in general keep a private log of all major activities such as an opposing player leaving the field or starting to perspire profusely (unless everyone is). This role of scorekeeper is underestimated by many. Doing it correctly and thoroughly is the mark of a winner.

In most sports, a good TRANSITION GAME is important if you want to win. This is also true in a negotiating game. Only a very few seconds are allowed in most sports to switch from offense to defense or vice versa. However, in negotiating you can have all the time you want—if you know how to get it; don't allow yourself to get trapped. When the other team makes an unexpected power play, either offense or defense, the best and only reasonable thing to do is to buy yourself some time before you respond. The amount of time you need depends on how extreme their move was and how critical the issue being resolved is to the entire game. Several ways to buy varying amounts of time are:

- You ask for a time-out to confer with your team.
- You pause indefinitely until the other team repeats the assertion or question or asks you to respond, to which you respond "I am thinking" and continue to pause.
- You ask the other team to repeat their stated position or probing question.
- You state that you do not understand their statement or question and want them to paraphrase it or expand upon it.
- You state that you are not prepared to respond right now but will do so at some stated or unstated time later in the game.
- You remind the other team that the proposition brought up is not in keeping with the approved agenda.

One final reminder on tactics. Not all tactics work in all games. Develop and practice your approach to using those we've mentioned and fine-tune those which you have created for the negotiating games that you've played before. No matter how good you think a tactic will be, always be willing to drop it or change to another one when circumstances dictate doing so. Remember, you can call an audible play at the line of scrimmage.

Chapter 13

Techniques for Effective Negotiating Communications

The essence of any negotiating game is complete and effective communication. Negotiation without it is like playing a ball game without a ball. Therefore, it is important for us to spend some time on techniques for TALKING TO, LISTENING TO, WATCHING, and PLAYACTING FOR the other team. Communication is employed by both teams to challenge, probe, explore, test, promise, concede, flatter, intimidate, demean, show appreciation, and even threaten. All of these uses and others that you can think of relate to one or more of the following four objectives of effective communication:

1. To discover the biases, prejudices, needs, preferences, and values of the other team.
2. To control or disguise the information given regarding your team's biases, prejudices, needs, preferences, and values.
3. To change in some way the relationship that currently exists between both teams, preferably for the better, but not always.
4. To influence the behavior patterns of the other team, both during *and* after the game.

Before we get into the specifics of how to communicate to achieve a winning game, there are two underlying principles that are the basic causes for poor communication, true for all human communication but particularly apt for the negotiating game.

> *First:* You assume or perceive that the other person or persons with whom you are communicating understand exactly what you mean or feel in what you said, without you asking them to tell you what they think you meant or felt.
>
> *Second:* You assume or perceive that you understand exactly what the other person or persons mean or feel in what they say, *without* you telling them what you think you heard or saw, or without asking them questions to make sure you understand.

If you could stop these two regular human practices, it would not be necessary to even lay out the techniques that are forthcoming. But, the reason that we regularly commit the first basic communications sin is because we perceive that if we question the other players as to whether or not they understood us, it will appear that we are questioning their intelligence, or manners, or trust, or whatever. We worry so much about the possibility of offending them that we would rather risk being misunderstood.

The reason that we regularly commit the second sin is because we don't want to make ourselves look bad, as though we are so dense that we have to ask questions or paraphrase their statement to make sure we understand.

The key to reaching maximum understanding is to stop worrying about your or the other person's ego and start asking questions or paraphrasing to validate your assumptions about what they said and your assumptions of how well they heard what you said.

Rules for Talking to the Other Team

A. *Select Your Words Carefully:* Certain words usually evoke an emotional response and move the two teams further apart. Always, should, don't, much, and never are just a few of the words that have a disciplinary, or scolding, or "I'm better than you" sound. It is also prudent to consider the connotations of certain words that may have offensive meaning to the other team's players. Example: "You should *never* bring up something like *that* in a negotiating game!"

B. *Control Your Judgment:* When the game gets heated up and intense, emotions can run high. At these times, judgments can be taken very personally. Most people deeply resent having their values, opinions, or competence judged by others. Describing how you feel about a situation or what you see usually is more productive than questioning or judging the other person's motives or line of reasoning. Example: "What you are asking for is *ridiculous!*"

C. *Use "I" Whenever Possible:* "I" is much more personal than the corporate "we" and sounds more collaborative. This is absolutely essential for one-on-one negotiations. In addition to staying away from "we" messages, also stay away from "you" messages, especially when "you" is followed by one of the absolute words given in A. Example: "*I* want to conduct this game so that both teams satisfy our needs."

D. *Regularly Compliment the Other Team:* A well-timed compliment may be worth an extra score in a tight game. Focusing on areas of agreement or throwing out a "you might have a point there" can break up the tension and make the opponents more willing to listen to your propositions more carefully. Example: "The information you've shared on this issue has certainly been helpful."

E. *Learn How to Ask Questions:* This may be the single most important technique in talking to the other team. Properly used questions are the best tools that experienced negotiators use. However, if they are used improperly or ineptly they can easily escalate the conflict and widen the differences on the issues being negotiated. As part of your game plan, on *every* issue to be resolved, your team should review and practice these three questions:

1. What questions should I ask?
2. How should I word or phrase each question?
3. What is the best time to ask each question?

Practicing the asking and answering of the right questions is one of the most important skills to develop during your planning sessions.

Open-ended questions that begin with what, when, where, how, or in-what-way get the most information from the other team. Closed-ended questions that must be answered with one or two words, usually yes or no, are very ineffective in soliciting expansive information. They can be used effectively, though, to bring closure on an issue.

You must be careful in using open-ended "why" questions. "Whys" can be very personal. Sometimes those who are questioned don't know why they feel like they do—they just do; or maybe they know why but don't want to tell you. "Why" questions generally make the other person feel defensive. A good approach to getting at "why" without using why is: "I'm sure you have several good reasons for saying that. Would you mind sharing some of them with me."

Questions can range from simple conversation openers like "How are you this morning?" to questions that create high anxiety and threaten the other person's self-image. When the objective is truly a winning game of negotiating, you should try to avoid all threatening, offensive, or punitive questions. Only extensive practice will enable you to accomplish this. Example of a threatening open-ended question: "What are your reasons that the project can't start until two months after the contract is signed?" Example of an appropriate closed-end question: "Do you agree that we have a deal if I make the change you just suggested?"

F. *Use Suspended Responses:* At any time in negotiating, but especially as you get into the tough issues when the dialogue becomes emotional, an effective tool is to suspend your response to a particularly harsh, emotional statement from the other team. Wait two to four seconds and with good eye contact and a firm (but not loud) voice say: "I feel your last response is not helping us make progress on this issue."

G. *Concentrate on the Current Game:* Experience has shown that it is seldom helpful in a negotiating game to dig up the past (especially to blame or demean the other team) or to speculate about what might happen in the future. Sometimes, events or issues may need some historical perspective, but concentrating on the current circumstances is usually best. Example to avoid: "I know you want those terms out of the contract because of the mess you made on the last job."

H. *Control Your Own Information Flow:* It is a good idea to practice being discreet and disciplined about your feelings toward the other team and about your own goals. Being brutally frank and honest can be more of a liability than an asset. We are not advocating a deceitful approach but something between insensitive candidness or blabbering and "beating around the bush." Example to avoid: "I'm going to have a hard time discussing these issues with you because you don't have as much education and experience on these projects as I do."

I. *Don't Use Overgeneralized Statements:* Generalizations tend to back people into a corner and make them react emotionally and defensively. Example: "You always underestimate your hours on project budgets."

J. *Do Not Interrupt Others When They Are Speaking:* When we interrupt, we are telling others that we don't think their ideas are worth listening to or what we have to say is much more important than what they have to say. This is usually just a bad habit that can only be broken by constant awareness.

K. *Avoid Excessive Talking:* This is another bad habit usually driven by ego. It totally discourages the other team from contributing and they will stop listening as you banter on and on. It might be good to point out that "when you are speaking you are only repeating what you already know and the only way for you to learn more (about the other team) is to let them speak and for you to listen."

L. *Don't Talk Down to the Other Team:* This approach sounds like a parent addressing a young child and most adults totally resent it. These kinds of comments suggest irresponsibility of the other person or team and represent a dominating approach. Example: "What kind of excuse are you going to use for our late start today?"

M. *Don't Use Loaded Questions:* Loaded questions leave no room for the other person to respond in his or her way. They're not questions at all but negative, demeaning statements that create tremendous resentment. Example:

47

"Have you ever considered sending us the material samples like we asked for in the specifications?"

N. *Don't Try to Be Cute:* Sarcastic, cynical, mocking comments will put the other team on a full defensive charge every time. Even when they are meant in jest, they will probably be perceived as an attempt to ridicule or to hurt. Humor can be used constructively as long as it is not aimed at an individual or any class of people. Be very careful in telling jokes. It is very easy to offend and set the game's progress back hours or days.

O. *Don't Use Foul Language:* The negotiating field is no place to show how many four-letter words you know. Even if you use them on the practice field, they should be totally avoided during the game. This applies even if the other team chooses to do so. Never say anything that you wouldn't want to hear repeated on instant replay. The same goes for off-color jokes. Why take the risk of offending someone and hurting your chances of a winning game.

A Few Pointers for Listening to the Other Team

We normally speak at an average rate of 150–200 wpm, and we can listen and comprehend in the range of 300–400 wpm. This leaves a lot of time for our mind to wander around not concentrating on what the other person is saying. This is one good reason why every person on the team shouldn't be trying to take detailed notes, especially if the game is being recorded. Besides, that is the scorekeeper's job. The following are tips to make you a completely active listener:

- Jot down key words and concepts as spoken by the other team.
- Be prepared to ask questions about what the other person just said or to use the brief notes to play back what you thought was said.
- Be comfortable in asking the other person to further explain what a statement or question means.
- Listen carefully for the feelings and intent behind the actual spoken words.
- Make eye contact regularly (but don't stare) at the speaker and look for gestures, facial expressions, and other body language to give you clues as to the emotions behind the words.
- Suppress your own positive or negative feelings until you fully understand the intent of what was said.
- Never be in a hurry to answer, no matter how great you think your response or question is.
- Be aware of your own facial and body movements in responding to what is

being said. Why telegraph your feelings unless it is a thought-out, well-planned tactical move.

For the experienced negotiator, active listening, using the above tips and any others that might be appropriate, is a tool that is as powerful as knowing how to ask questions. For further help, see Appendix E.

Tips on How to Watch the Other Team

Most of the communication in the negotiating game which we have thus far discussed relates to using our senses of speaking and hearing. The third sense that can play a major role in the game is eyesight. An interesting fact about our eyesight is that when used as a receiver of information it is much more effective than our ears. As we have pointed out, we can hear and determine the meaning of words at an average of 400 wpm, maybe even up to 500–600 wpm, depending on the speaker and content. However, if we use our eyes for one minute to observe a busy meeting room or a crowded street filled with people and then tell someone or write down what we saw, it can easily take more than 1,000 words to do so.

Thus, we should learn to be more visually aware of what is going on around us in the negotiating game. This is particularly true of the movements of the other team's players. These movements are commonly called non-verbal communication or body language. It involves the posture, facial expressions, hand and arm gestures, and other body movements that are voluntary and involuntary. Some communication authorities suggest that up to two-thirds of the total information conveyed in a one-on-one conversation is conveyed non-verbally and only one-third verbally.

Body language then becomes particularly important in negotiating games because of the tremendous quantity of informational cues that are transmitted by the participants. To understand what these cues are and what they mean becomes important to any negotiator for two reasons:

1. To be able to pick up and understand the cues emanating from one's opponents.
2. To be able to influence, perhaps in some cases even mask, the involuntary cues emanating from yourself or your teammates.

Most authorities believe facial expressions most directly reflect the sender's emotions. Six different emotions can be clearly identified through facial expressions: happiness, sadness, surprise, fear, anger, disgust. The expressions may reflect true emotion and may accurately accompany the verbal communication.

On the other hand, facial expressions as with other body language, may contradict the spoken word, thus sending out the real message.

Body movements tend to reflect attitudes rather than emotion, such as liking and disliking. Liking someone is often characterized by a more forward lean, a closer proximity, more eye gaze, more openness of arms and body, and more postural relaxation than disliking. The use of hands on hips by a standing communicator seems to be used more frequently when the attitude is one of dislike rather than liking.

One should be cautioned, however, against relying on any single cue. Different cues may be more or less helpful depending on conditions. For example, in attempting to determine whether a person is outgoing or reserved, cues from the face seem to be most useful. However, in determining if others are relaxed or tense, body cues may be more meaningful. In short, there is no single source of non-verbal information for all situations.

See Appendix F for more extensive tips and guidelines on the use of body language and the observation of it.

Tactics for a Game Out of Your League

If you get stuck in a game with a very powerful opponent and he or she doesn't want to play a game of winning negotiation—watch out. Let's face it, no technique or strategy can guarantee a winning outcome if the other side has all the power and is determined to use it to the fullest. Realities being what they are, all of one's skills may not be able to totally counteract such an opponent. However, things aren't always as bad as they often seem to be. You may have more going for you than you realize. Rarely is one totally powerless. Usually feelings of powerlessness come from untested assumptions. With reflection and planning you can at least protect yourself from being overwhelmed, that is, being forced into accepting an outcome that will be regretted far into the future.

When faced with such odds, it becomes imperative for you to examine in depth your acceptable target range on every issue, especially major ones. Be sure they are accurate. Be sure you understand what alternatives you have to these "bottom lines." What happens, for example, if you don't reach a settlement? The better the alternatives, the less vulnerable you will be. Spend time exploring and developing these alternatives.

If the other side has overwhelming physical or financial resources, one should appeal to their sense of principle and independent standards of fairness. Having a sound alternative to your acceptable target range will help you accomplish this.

Tactics for an Uncooperative Opponent

Frequently, business and personal situations arise where we are forced into a game with an opponent who doesn't share our winning game philosophy. The other team quickly states their positions on all issues and then refuses to budge. The temptation is to walk away and you should, unless you have needs that the opponent *possibly* can meet. As long as the opponent is willing to talk, even a little bit, there is a chance for a winning game.

Whatever you do, don't criticize your opponent's game plan, BUT stick to your own. It is tempting to criticize a clearly unfair position taken by an opponent. To do so, however, is playing the other team's game. The criticism will normally elicit defensiveness, even causing the other team to become more militant and unreasonable. Likewise, when an opponent attacks your proposals, there is a natural tendency to defend them. It is suggested that you neither criticize your opponent's position nor immediately defend your own. Instead, suggest that both parties examine the underlying reasons for such positions, thus getting to the true interests and needs.

The basic principles of a winning game of negotiating are often contagious if one has enough patience and persistence. Approached in this manner, many uncooperative negotiators will eventually see the merits of a mutual exploration of needs rather than a clash of positions.

Many may see the merits of such an approach but some won't. So if a settlement truly needs to be made and all else fails, consider suggesting an impartial third party for some form of Alternative Dispute Resolution (ADR). Such an individual should be acceptable to both parties in advance and be the kind of individual or panel of experts who can effectively mediate such a dispute.

An Assortment of Trick Plays and Traps

There may be times when you will have to use techniques that are normally not consistent with a winning game of negotiating. It is usually best not to start the game using these but only do so after the other team shows that it is ignoring the principles of a winning game. This means that the other team has started using it's own trick plays, traps, or resistance defense. So, even if your intent is not to use these plays on offense, you need to know and recognize them in order to defend against them. Several of the nastier trick plays and traps are:

A. *Anger:* Some opponents will "lose it" on purpose to test you. Anger or emotional tirades by an adversary can either be real or faked. If in doubt, it's best to assume it's real. Anger usually occurs when an individual per-

ceives an external event (object or person) as threatening or intimidating or when we experience the frustration of unmet expectations. Both situations can occur during negotiations. Anger from the other team offers a high potential of hooking you into a cycle of compounding anger as well; this increases the difficulty of winning negotiations. The following steps may be helpful in dealing with anger in the opponent:

- **Affirm the other's feelings**—to deny, to ridicule, to minimize, or to disallow the other person's feelings usually magnifies them.

- **Acknowledge your own defensiveness**—admit your own tenseness. Say something like "you know I'm feeling a lot of emotion, too. Let's take a break and try to relax a few minutes."

- **Clarify the situation**—use the active listening skills discussed earlier. Determine exactly what the threats or unmet expectations are.

- **Re-establish the relationship**—apologize if you said something demeaning or intimidating (even if you don't think you did). This will not hurt your position if you are truly playing a winning game.

B. *"I Know You Can Do Better Than That"*: Often used by a buyer in a straight buy-sell negotiation to tip the balance more favorably toward the buyer. This frequently puts the seller on the defensive and can compromise the no-lose objectives. Since this is an extremely general statement, a good response is to get back to the specifics that will satisfy the needs of both parties. Example: "Mr. Buyer, are you referring to price only or to some other issues that are involved. I'd like to deal with the specifics of what you mean."

C. *Mixing Real Issues With Straw Issues*: In a multi-issue negotiation, one trap that an opponent may use is the real vs. straw issue tactic. This involves the other team introducing into the negotiations several real issues together with several marginal or low-priority issues but pretending they are all equally important. The tactic for the opposition then is to make generous concessions on the straw issues in exchange for major concessions from your side on the real issues. The overall appearance to the inexperienced negotiator may seem to be a balanced set of concessions while in reality very little of importance has been conceded by the adversary. Defense of this play was covered more thoroughly in "Using the Concession Stand." (See Chapter 12.)

D. *Sandbagging:* Sandbagging is a form of deception. It is designed to lure the intended victim into a false sense of security—but only for a while because sooner or later the perpetrator intends to cash in on the victim's complacency or vulnerability. In the negotiating game, it shows up as apathy in the other team to an issue or the agreement on that issue. In reality, the opponent is very much against the issue or proposed agreement and plans to cram it down your throat later. (This is a classic slam dunk.)

E. *"Deadlocked":* The deadlock is one of the more powerful traps in negotiations. The opponent's position may be sincere or it may be a bluff but, either way, it threatens a winning game. Here are some thoughts on how to break a deadlock:

- **Suggest a time-out**—Change the location and perhaps the time to a less threatening location. Allow emotions to level off.
- **Ask for time to get more information**—Again, allow emotions to cool.
- **Rephrase your proposal**—Perhaps the opponent didn't understand all the points the way they were intended.
- **Mini-concessions**—Suggest to your opponents that each team make a very small concession to break the ice and get things moving again.
- Bring in a neutral **third party mediator** to clarify true interests and suggest alternative approaches to a winning solution.

See Appendix C for additional suggestions.

F. *"I Don't Have The Authority":* Whenever the other team is asked to make a concession, the response is "I must discuss your proposal with my boss." This tactic, in effect, creates a one-way flow of concessions. The opponent is quite willing to accept any concession you may offer to make but doesn't have the authority to personally reciprocate such a concession. The best way to defend against this trick play is to always get authority-level issues cleared up at the start of the game.

G. *The "Classic" Trap:* The classic trap play is a form of deception similar to sandbagging. The opponent attempts to fool you by creating an illusory weakness. Your team, attempting to capitalize on the weakness, pursues it with considerable vigor and when fully committed, the trap is sprung.

H. *Loaf of Bread Trick:* This involves approaching an issue a slice at a time and accomplishing one's entire objective over a period of time. This approach is often used to get one's foot in the door when seeking new business, but it can be used in any negotiating game.

I. *Piggybacking:* Once agreement is reached on the major issues an attempt is made by the other team to quickly add one or more minor points, on the assumption that they will seem insignificant by comparison.

J. *The "I Already Know" Lie:* This involves one party intentionally presenting a factually incorrect statement as true in order to gauge the opponent's reaction. It is often used as a means of obtaining information without divulging any. Example: "I know that your company paid $50 million for the last plant like this that you built." "Why sir, I don't know where you got your information, but we only paid $44 million."

K. *Split the Difference:* This seems so fair but can really work against you; success depends on where each team started from on the issue and if the compromise point is in your acceptable target range.

L. *Flat-Out Lie:* This trick play involves the written or verbal falsification of documents to mislead the other team on such things as costs, budgets, bids, etc. This play is used to trick you into making a significant concession in the negotiation. This negative tactic is used frequently by companies to "shop" contractor bids.

M. *A Fait Accompli:* As the old saying goes, "it is better to ask for forgiveness afterwards than to ask for permission before." When the other team goes ahead and carries out an action rather than negotiate with you on it, you are being told "the act is done, what are you going to do about it?" This is the total antithesis of a winning game of negotiating.

N. *Good Guy–Bad Guy:* This is a tactic used by two people on the same side against an opponent. One of them plays the role of the bad guy, usually making abrasive and unreasonable demands, perhaps pushing the adversary to the limit. The good guy, on the other hand, maintains a warm sympathetic liaison with the opponent in anticipation of future dealings.

O. *"I Just Can't Afford It (You)":* In this tactic, the buyer doesn't question the validity of the asking price but merely states that there is not enough money available to meet that price. This tactic is usually accompanied by various ego-building comments.

P. *Fool's Gold or Salting the Mine:* This approach is used to carefully and thoroughly make an issue or agreement appear more attractive or lucrative than it actually is, by deception of all kinds.

The better you understand each of these trick plays or traps, the better will be your defense against each one. Sometimes the very best defense is to just acknowledge the trick play or trap by naming it. However, don't criticize the other team for using it. Just letting them know you know should be enough to get them to stop.

Part 4
On Reaching Closure:
Tallying the Final Score and
Mounting Your Trophy

Chapter 14

As the Clock Winds Down

There are two primary ways for a winning game of negotiation to end: all of the issues have been resolved satisfactorily for both teams or you hit a time deadline. Either or both of these can be somewhat arbitrarily determined at the beginning of the game. If you find that the pre-set deadline is fast approaching and there are substantive issues undiscussed or unresolved, you have three choices: go into overtime, suspend the game until a new date can be set, or consider the game complete, regardless of what the scoreboard reads. The last choice is unthinkable in a winning game of negotiating and so it remains for the two teams to decide on overtime or suspension. The needs of both teams and all starting players should be considered in deciding which is best. Factors that may help in the decision are:

- The mental and physical condition of both teams.
- Urgent commitments by key players of either team that can't be rearranged.
- The degree of cooperative spirit currently existing in the game.
- The need for one or both teams to do further research or analysis before the last few plays.

Both teams should consider these and any other applicable factors and make a joint decision. *One team may try to use the deadline, as mentioned before, as a power base to extract large concessions in the short time remaining. Do not fall prey to this!* The game is not over until both teams agree that it's over and too much time and energy have probably been invested for either team to consider forfeiting at this late inning.

Also, don't fall for the end of the game trap where the other team says, "Oh, we only have two to three minor issues left. We're real close on those. We'll just write it up and send it to you. There shouldn't be any problem with you agreeing." If the issues are that minor (which they probably aren't) and you're that close to agreeing (again, probably not), then just ask for a short overtime to resolve them together. Sometimes bully teams will try to run in some last minute substitutions to confuse your offensive and defensive plays. First, call a foul—the only game in which a player can call a foul—and then, a long time-out. If you accept substitutions for a few plays before calling a foul and a time-out, you've lost your momentum and leverage to stop an illegal substitution.

In major league games, it's always a good idea to call for a time-out near the end, even when there are no major time problems or trick plays being thrown at you. The purpose of this final time-out is to get your team together and carefully check the scoreboard. Make sure you understand how the scoreboard reads on all issues and make a last minute run to the concession stand to see if there are any late-in-the-game treats that you've forgotten about, to help in your final scoring drive. This last time-out can be one of the most important ones in the entire game.

As the issues and time come down to the wire, there are a few last-minute plays to use in wrapping up the game:

- Make controlled and creative final offers.
- Skillfully question the reluctance of the other team to move on an issue (a good place to use those treats you picked up at the last time-out).
- Appropriately assert your expectations for specific concessions from the other team.
- Directly ask for final agreement.
- Express a positive attitude toward all of the work that has been accomplished.
- Check the official scoreboard together with the other team to make sure both sides agree on the individual as well as the final scores (both teams should think they won).

This last point can't be over-emphasized. As previously suggested, each team will probably have its own private scorekeeper and those notes and records are very important. However, the official scoreboard should be the one recorded for both teams on an easel or blackboard. This recap of all the scoring can be done at the end of the game or, for lengthy negotiations, at the end of each day or after every major issue or resolution. If it has been a lengthy process, it is actually better to not wait until the end to review the entire game. The other team may use this time to nit-pick on an earlier agreement or rearrange the scoring from a much earlier scoring play by your team.

Another reason to go over all the scoring before the teams leave the playing field is to reduce the likelihood of you receiving a call the next day from the other coach to say, "There was one small item I forgot to mention on issue so and so. I know you won't mind adding it to the agreement." If this does happen before the game is documented for the record books and ready for signing, just remember that if you agree to this small addition you may get hit with a few more of them. Your choices are to say no, say yes but with a reciprocal concession on the other team's part, or say yes without anything in return. These are in order of preference and there should be a pretty good reason to even go to choice number two. "No" with some nice words attached is the best by far.

So in summary, do not leave the playing field until both teams have reviewed all scoring to confirm all agreements reached on all issues. If reviews have been held regularly throughout the game, then a review of major issue agreements may suffice at the end. Only after this review is complete should you allow the final whistle to blow and the gun to go off. Then the game is over (for now). Now commences the management and execution of whatever was agreed upon. "Ladies and gentlemen—Start your engines!"

Chapter 15

Post-Game Niceties

Don't clear the field too fast. The attitudes that have contributed to a winning game of negotiation should carry over to the post-game activities while both teams are present. Following is a list of do's and don'ts for this critical period:

- Do shake hands and cordially chat with the other coach at mid-field.
- Do plan ahead for all the other players on your team to do likewise with the other team's players: remember, both teams feel that they have won.
- Do agree as to who will accept the responsibility for the final preparation of the game record, timing to do so, review by both teams, and process to make it official (especially important if the game result is a contract and persons other than those negotiating need to sign the final document).
- Do agree with the other team in what form the written documentation will be:
    a. Memo—a summary of what was accomplished.
    b. Understanding—an outline of the agreed-upon common interest of both teams.
    c.. Agreement—an official (and almost always signed) document that commits both teams to all of the detailed terms that have been negotiated along with the specific responsibilities and expectations of both teams.
- Do agree on the final outcome as it will be given to the owners. After some games, both coaches may want to talk to reporters to give their assessment of the game. In other cases, there may be good reasons to decide that neither coach will talk to the press and it will be decided by the owners what, if anything, will be released to the public domain. Major negotiating games do not always create a need for major press conferences, depending on the confidential nature of the game.
- Do thank the other team for the professional manner in which they played the game (even if they tried a few unusual plays—you probably did too).
- Do tell the other team that you look forward to playing more negotiating games with them.
- Do not start any joyous celebrating, no matter how great you played or what the final (perceived) score was, until your team has completely left the playing area. Loose, indiscreet tongues have killed many good games after the final whistle was blown and while the other team was still present.

- Do not boast or complain about any issue, major or minor, that was decided during the game. One wrong comment may cause the other team to reconsider or at least invite them to look at their "winning" differently.
- Do make sure you collect all of your private notes, research, scorekeeping, etc.
- Do weigh carefully the pros and cons of having a social affair with the other team to celebrate. Celebrations are usually best held after the game has officially been accepted by the owners. Social occasions still require discreet communication, before and after the outcome is official, especially if you want the next game to be a winning one also.
- Do not *ever* say to anyone publicly or privately that the other team did not win in order to make yourself look good. Even if your team scored better than anticipated, if you played the game correctly, the other team should feel like they won (too). So, don't blow it (and the chance for future games with them) by gloating and ruining a good winning game. If the other team does, don't reciprocate.

Chapter 16

The Post-Game Wrap-Up

In very few winning games does either team get everything it deserves or desires to the fullest extent. That's exactly the reason that we spent so much time on the game plan and developing the acceptable target range for all issues. Therefore, one of the most important procedures that all professional teams follow is to review the game films with the entire team as soon as practical. That may be immediately after the game, the next day or when you arrive back at your home field. The longer you delay doing this, the less impact it will have for your next practice and next game, even if you use a different starting team. In many instances, it is very educational to go over the film with the team members who weren't on the traveling squad for this game so that they too can benefit from your experience.

This is not a five-minute review on "fast forward." Done properly, it involves looking at your entire game plan and comparing the actual game to it. Deviations from your offensive and defensive strategies are particularly critical but so is determining which tactical plays were effective and why. Notes from the game itself can tell you which questions were particularly effective in getting information from the other team. Questions that created insights into feelings, interests and true needs are vital to remember for the next game—with this team or others. Learn from the body language signals that team members picked up so that you will all be more alert to them next time. If there was one scoring drive by the other team that caught you a little unprepared, brainstorm alternatives for defending against that play the next time any team tries to use it.

Another thing to keep in mind before you do much celebrating. A game does not insure a winning season. Almost all negotiating games are followed by smaller negotiating games that are required to follow-up on and implement what both teams agreed to in the original game. Some even say that when you've finalized a major agreement, the truly big game of negotiating has just started. Others say that completing a contract negotiation is just the rehearsal for the real game. Whatever your negotiating game is about, both teams must live up to the final score.

Part 5
Away Games

Chapter 17

International Negotiations

International negotiations are a whole new game. Nevertheless, most of the principles listed in this book will be applicable but you must remember that this will be a "foreign" experience!

Let us assume that your firm has been selected for a project in Hungary. You have been aware of the potential project and even had some input to the proposal process. Now you have the opportunity to participate in the negotiation. What should you do to get prepared as a member of the negotiating team? As a case in point, let's go through the process step by step.

- Learn as much as you can about the project by reviewing pertinent correspondence, the request for proposal, your firm's submittal, selection notification, proposed draft contract (if none, develop your own), and the (project's) funding source.
- Do research on the country, its people and culture, contracting traditions and laws (see notes at end of this chapter for some references).
- Determine the schedule for negotiations and who the participants are going to be. (It is assumed that the negotiations will be conducted at the client's office in Budapest.)
- Find out the names, titles, and functions of the client's negotiating team. Determine if they had been involved in the proposal and selection process. Try to find out their professional as well as personal background.
- Before you leave for the trip, do as much homework as you possibly can. Practice role playing with your team before you leave. (It is much easier to discover that you need something else before you are 6,000 miles from the home office.) Make notes, identify areas of concern, list questions you hope they don't ask, have an idea of your walk-away price, or walk-away position on other issues, list the pros and cons of obtaining this contract, and determine if you will be fully committed to this project after contract award.
- Leave a few days before the commencement of negotiations. It is recommended that you fly business class directly to your final destination. Schedule at *least two and a half days* before the first meeting. During this time rest, relax, adjust to the new time zone, and start final preparations with your team. If possible, visit the site of the negotiations beforehand.
- Keep in mind the negotiating strategies and tactics that have been described previously: e.g., logistics, agenda, ground rules, etc.

- Patience is a virtue. It is even more of a must at a foreign negotiating table.
- Don't expect a quick deal. It will take time and you better plan accordingly.
- Be focused and don't think about what's going on back at the home office. If at all possible, you should not be distracted in any way.
- If things get tough—and they will—take a recess of a day or so to collect your thoughts and determine new strategies and tactics.
- Make recaps daily and have notes prepared for the next meeting. These should summarize agreements reached and outstanding issues.
- Be persistent in a diplomatic way.
- Foster relationships.
- Be firm in your convictions.
- Don't become intimidated by the client's representatives.
- Remember it is a PROCESS!
- Don't be afraid to ask for assistance from the home office. Keep them apprised of the major issues and concerns.
- Get ample rest at night and do not overindulge in strange food and drinks.
- Do your best and be prepared to walk away from the contract if the client has unreasonable demands. Remember it must be a WIN-WIN situation.

Notes:

A. *U.S. Government Publications*
- *Country Marketing Plan — (country)*
- *Doing Business in (country)*
- *Overseas Business Reports*

The above publications can be obtained from:

Superintendent of Documents
U.S. Government Printing Office
Washington, DC 20402

Further information can be obtained from:

Director General
United States & Foreign Commercial Service
International Trade Administration
U.S. Department of Commerce
14th & Constitution/Room 3802
Washington, DC 20230

B. Country Culturegrams

A four- to six-page brochure series on the customs, manners, and lifestyles (as well as other applicable travel information) of more than 100 countries. The brochures are updated frequently, and are available individually or as a set at the address given below. The series is an offshoot of the Brigham Young University support of the Mormon worldwide missionary program. Very valuable for those in the international marketing area or charged with keeping employees behaving in a culturally appropriate manner.

David M. Kennedy International Center
Brigham Young University
Provo, UT 84602

Chapter 18

U.S. Government Contracting

Doing business with the U.S. Government can be a profitable or a costly experience. Whether a company makes money, loses money, or breaks even depends to a great extent on the understanding of the contract which it signs. Negotiations play a major role in the determination of a fair and reasonable price for the professional services required. Unlike those with the private sector, the negotiations with the government sector are more complex because of all the rules and regulations that must be complied with. Nevertheless, it can be a rewarding experience if you know the basics, rules, regulations, and environment you are dealing in. In addition, you must effectively manage your contract/project and have the courage to stand up for your rights.

Let's Get Down to Some Basics

- What type of contract is it? (e.g., fixed-price or cost-reimbursable)
- Are you aware that there are many variations of fixed-price and cost-reimbursable contracts?
- Do you *and* your client really know the scope of work?
- Is the scope of work vague, loose and/or unclear? If fixed price, is the scope of work clear and very specific?
- Do you know what forms and reports are required? Do you have copies of same?
- Do you understand the government jargon?
- Do you have a copy of the government regulations that will govern your contract? Do you know what they mean?
- What are the cost principles that are applicable to your contract?
- Do you know of TINA? (Truth in Negotiations Act)
- Are you aware of the audit provisions?
- Are the schedule and budget realistic?
- Are you willing to perform only the services that are authorized under the contract? (In other words, you are not going to give away free work.)
- Are you willing to do extra work on fixed-price projects only after negotiating and finalizing the change order?
- Do you know how to use cost data in your negotiations?
- Do you know the limitations on profit?

Some Areas That Must Be Negotiated and/or Approved

- The scope of work, schedule, price, terms, and conditions of the contract.
- Adjustments with regard to the furnishing of Government Property and Facilities.
- Responsibilities, obligations, and liabilities of the parties.
- Overhead rates for cost-type contracts.
- Acceptability of accounting and purchasing systems.
- The approval of subcontracts.
- Alterations to contract clauses.
- The amount of profit to be earned (under fixed-price and cost-reimbursable contracts).
- Payment and retention provisions.

Process

The company (also known as the Contractor), during the term of a Government contract, may deal with the Contracting Officer, Cost Analyst, Legal Staff, Technical Personnel, Auditors, Property Administrators, Security Representatives, and other Government Personnel concerned with the performance and administration of the contract.

The company, therefore, is not dealing with the "Government" in the procurement process. It is dealing with the Contracting Officer and the Contracting Officer's representatives. These representatives are individuals that have virtues and vices that other people have. As we stated before, there are many detailed regulations covering the Government procurement process. However, they only establish the broad limits within which the Contracting Officer and the Government Representatives must operate. Within the regulations, the Government personnel have wide latitude in which they may exercise judgement. Therefore, they call them as they see them; and you have the right to question their decisions.

Recap

Negotiating with the government is a challenging experience because you are dealing with people who do that for a living. You will find that they are well trained, have more contracting experience, and are always looking for the best contract. They are assisted by in-house experts (e.g., auditors, cost analysts, legal, and technical staff) that enhance their position.

In order to be successful with your government contracts you must be prepared to meet them on a level playing field. If you are not well versed in the regulations, you are at a significant disadvantage.

The following are some rules professionals should follow when negotiating with the government:

- Learn the true negotiation process and recognize it could take days, weeks, or months to consummate a contract.
- Work in a team environment. When negotiating, include a principal and someone who can address the project's technical issues. If possible, include the proposed project manager. Also, bring along someone who's familiar with government regulations and has a financial background.
- Don't bring your lawyer unless the other side does. If you bring a lawyer and the government doesn't, you're setting up a hostile situation. However, have your lawyer review the contract once it's finalized and before it's executed. Also, have your lawyer available to answer legal and liability issues.

Remember, when negotiating with the government, hope for the BEST and prepare for the WORST.

Part 6—Appendices
Training for Your Next Negotiation

Appendices

A. Checklist for The Ideal Negotiator 77
B. Ten Primary Reasons for Failure in Negotiation 78
C. Checklist on How to Break Contract Deadlocks 79
D. Checklist for Your Negotiating Effectiveness 80
E. Checklist on How to Improve Listening Skills 81
F. Attitudes Communicated Non-Verbally 82
G. Recommended Reading 85
H. Negotiation Case Study 87

Appendix A

Checklist for The Ideal Negotiator

1. Ability to negotiate effectively within one's own organization, family or circle of friends.
2. Willingness and commitment to plan carefully and know the product or service. Courage to probe and check information.
3. Good business judgment. Ability to discern real bottom-line issues.
4. Ability to tolerate conflict and ambiguity.
5. Courage to commit to higher targets and take the risks that go with it.
6. Wisdom to be patient, to wait for the story to unfold.
7. Willingness to get involved with the opponent and the people in his or her organization.
8. Commitment to integrity and mutual satisfaction.
9. Ability to listen with an open mind.
10. Insight to view negotiation from a personal standpoint, to see hidden personal issues that affect outcome.
11. Willing to use team experts.
12. Ability to negotiate with oneself and laugh a little, without too strong a need to be liked—high self-esteem.
13. Ability to clearly articulate one's ideas, desires, opinions or positions on issues.
14. Stability—emotionally controlled under high stress.
15. Keen ability to read people.

Appendix B

Ten Primary Reasons for Failure in Negotiation

1. Not refuting invalid assumptions; poor listening.
2. Asking few or poor questions; asking them at the wrong time.
3. Both parties making few meaningful disclosures.
4. Separating fact-finding from negotiating.
5. Not resolving semantic difficulties.
6. Closing the door on alternatives.
7. Allowing prospective problems to negatively affect the negotiator.
8. Not enough recapping of what has been agreed upon.
9. Establishing limits too quickly.
10. Not effectively using all team members.

Appendix C

Checklist on How to Break Contract Deadlocks

Deadlocks are common to all negotiating. You can break deadlocks by:

1. Changing the shape of money (not the amount): for example, advance payments, shorter payment terms, wire transfers, etc.
2. Postponing difficult portions of an agreement for negotiation later.
3. Sharing risks.
4. Changing the atmosphere from competitive to cooperative.
5. Changing the base for a percentage. Smaller percentage of larger base or larger piece of smaller base.
6. Adding options.

Appendix D

Checklist for Your Negotiating Effectiveness

How effective are you in negotiations? Rate yourself between 1 and 5 on each of the following items. Never—1, Sometimes—2, Frequently—3, Almost Always—4, Always—5.

	Self-rating
Knowing what you want	_____
Knowing what you are likely to get	_____
Knowing what the other party really wants	_____
Knowing what the other party will give	_____
Having necessary facts and information	_____
Planning for your presentation	_____
Anticipating the other party's presentation	_____
Presenting your position persuasively	_____
Listening openly	_____
Defending your position assertively	_____
Trading—getting and giving concessions	_____
Planning appropriate questions	_____
Using your power	_____
Developing alternatives	_____
Gaining respect and trust	_____
Avoiding pitfalls	_____
Being aware of critical timing	_____
Reaching clarity of agreement	_____
Overall negotiating effectiveness with:	
Individuals	_____
Groups	_____

If you would like to have your negotiating effectiveness evaluated subjectively by the authors, make a copy of your scores and send to O.C. Tirella, 11573 San Juan Range Road, Littleton, CO 80127.

Appendix E

Checklist on How to Improve Listening Skills

1. What are your particular strengths or weaknesses as a listener?
2. What factors tend to distract and take your mind off the speaker's subject?
3. Are there any mannerisms or gestures which cause you to "turn off" the speaker?
4. What words, phrases, or illustrations might cause you to react negatively towards a speaker?
5. The next time you hear a speaker, pay particular attention to his or her opening statement and closing.
6. The more attention you pay to nonverbal communication, the better listener you can become due to your increased awareness of the other person.
7. Be prepared to ask speaker to paraphrase or restate if you do not understand or restate what you thought you heard and ask speaker if that is what was meant.
8. Take time to listen. Whenever you sense that someone is troubled, about to "blow his stack," or needs to talk, give them the floor. Though it may seem like a waste of time to you, it seldom is.
9. Don't overreact. If an emotional volcano erupts, your best response will be to let it flow uninterrupted until it is exhausted. Make every effort possible to understand what is said and make the speaker feel as though it is important.
10. Be patient with your curiosity. There's a distinct difference between willingness to listen and curious inquisitiveness designed to obtain information. Sometimes in listening one can learn more about a situation by just listening than by probing which often causes the other person to withdraw.
11. Control your urge to be judgmental. You should refrain from instantly evaluating what is being said.
12. Give advice very carefully. Even persons who want to emotionally express their feelings may sometimes ask for advice. Be wary of giving any.
13. Don't underestimate the speaker's ability to solve his or her own problems or come up with a creative alternative. As individuals speak, they are really thinking things over. If you refrain from speaking, the chances are fairly good that they will work things out for themselves.

Appendix F
Attitudes Communicated Non-Verbally

Openness	Open hands Unbuttoned coat
Defensiveness	Arms crossed on chest Legs over chair arm while seated Sitting in armless chair reversed Crossing legs Fistlike gestures Point index finger Karate chops
Evaluation	Hand to face gestures Head tilted Stroking chin Peering over glasses Taking glasses off—cleaning Glass earpiece in mouth Pipe smoker gestures Getting up from table—walking around Putting hand to bridge of nose
Suspicion	Not looking at you Arms crossed Moving away from you Sideway glance Touch/rub nose Rubbing eye(s) Buttoning coat—drawing away
Readiness	Hands on hips Hands on mid-thigh when seated Sitting on edge of chair Arms spread gripping edge of table / desk Moving closer Sprinters position

Reassurance	Touching Pinching flesh Chewing pen/pencil Biting fingernails Hands in pockets
Cooperation	Sprinters position Open hands Sitting on edge of chair Unbuttoning coat Tilted head
Confidence	Steepling Hands in back—authority position Back stiffened Thumbs in coat pockets with thumbs out Hands on lapels of coat
Territorial dominance	Feet on desk Feet on chair Placing object in a desired space Elevating oneself Cigar smokers Hands behind head—leaning back
Self-control	Holding arm behind one's back Gripping wrist Locked ankles Clenched hands
Nervousness	Clearing throat Whew sound Cigarette smokers Picking-pinching flesh Fidgeting in chair Hands covering mouth while speaking Not looking at the other person Jingling money in pockets Tugging at ear Perspiration/wringing of hands

Frustration	Short breaths
	Tsk sound
	Tightly clenched hands
	Wringing hands
	Fistlike gestures
	Pointing index finger
	Running hand through hair
	Rubbing back of the neck
	Kicking at ground or imaginary object
Boredom	Doodling
	Drumming
	Legs crossed—foot kicking
	Head in palm of hand(s)
	Blank stare
Acceptance	Hand to chest
	Open arms and hands
	Touching gestures
	Moving closer to another
	Sitting on one leg
Expectancy	Rubbing palms
	Jingling money
	Crossed fingers
	Moving closer

Appendix G

Recommended Reading

*5 *The Art of Negotiating*, G. I. Nierenberg, Cornerstone Library, New York, N.Y., 1968

Body Language, Julius Fast, Evans & Company, New York, N.Y., 1970

Body Talk, Maude Poiret, Award Books, New York, N.Y., 1971

Conflict and Cooperation in Management (HBR Reprint Series No. 21059), Harvard Business Review Press, Boston, Mass., 1963–1973

Conflict Regulation, Paul Wehr, Westview Press, Boulder, Colo., 1979

Creative Aggression, George R. Bach and Herb Goldberg, Avon Press, New York, N.Y.,1975

Dealing With "No," Houghton Mifflin, Boston, Mass.

Effective Conflict Management, Roy Pneuman and Margaret Bruehl, Prentice-Hall, Englewood Cliffs, N.J., 1982

Fundamentals of Negotiating, G. I. Nierenberg, E. P. Dutton, New York, N.Y., 1977

*6 *Games People Play*, Eric Berne, Grove Press, New York, N.Y., 1964

*3 *Getting to Yes*, Roger Fisher and William Ury, Houghton Mifflin, Boston, Mass., 1981

Give & Take: The Complete Guide to Negotiating, Strategies, and Tactics, Chester Karras, HarperCollins, New York, N.Y., 1974

*4 *How to Read a Person Like a Book*, G. I. Nierenberg and H. H. Calero, Hawthorn Books, New York, N.Y., 1971

*10 *How to Win Friends and Influence People*, Dale Carnegie, Simon & Schuster, New York, N.Y., 1981

Human Behavior at Work, Keith Davis, McGraw-Hill, New York, N.Y., 1971

I'm O.K.—You're O.K., Thomas Harris, Avon Books, New York, N.Y., 1971

Inside Intuition, Flora Davis, Signet, New York, N.Y., 1971

Interpersonal Conflict Resolution, Allan C. Filley, Scott, Foresman & Co., Glenview, Ill., 1975

Interpersonal Peacemaking: Confrontations and Third Party Consultation, Richard E. Walton, Addison-Wesley Publishing Co., Reading, Mass., 1969

Managing Inter-Group Conflict in Industry, R. R. Blake, Gulf Publishing Co., Houston, Tex., 1965

Mannerisms of Speech & Gestures in Everyday Life, Sandor S. Feldman, International Universities Press, Madison, Conn., 1959

Meta-Talk, G. I. Nierenberg and H. H. Calero, Simon + Schuster, New York, N.Y., 1974

Negotiating Game: How to Get What you Want, Chester Karras, HarperCollins, New York, N.Y., 1970

Negotiating Higher Design Fees, Frank A. Stasiowski, Whitney Library of Design, New York, N.Y., 1985

Negotiating Rationally, Max H. Bazererman and Margaret A. Neale, The Free Press, New York, N.Y., 1991

[*1] *Negotiator—A Manual For Winners*, Royce A. Coffin, American Management Association, New York, N.Y., 1973

New Approaches to Conflict Resolution, Ford Foundation, Ford Foundation Office of Reports, New York, N.Y., 1978

New Ways of Managing Conflict, Rensis Likert and Jane Gibson Likert, McGraw-Hill Book Co., New York, N.Y., 1976

[*7] *Power: How to Get It, How to Use It*, Michael Korda, Warner Books, New York, N.Y., 1981

The Japanese Negotiator, Robert M. March, Kodansha International, Tokyo, Japan, 1988

The Silent Language, E.T. Hall, Fawcett Publications, New York, N.Y., 1959

[*8] *Winning the Negotiation*, H. H. Calero, NAL Dutton, New York, N.Y., 1979

[*9] *Winning Through Intimidation*, Robert J. Ringer, Fawcett, New York, N.Y., 1984

[*2] *You Can Negotiate Anything*, Herb Cohen, Lyle Stuart Inc., New York, N.Y., 1980

---

*Authors' suggested priority reading order.

Appendix H

Negotiation Case Study

Facts:

1. In March of 1972, a St. Louis engineering firm (Firm) was selected for a project by the city of Knoxville, TN (City). Since the schedule was tight, the City and the Firm entered into a one page letter contract (L/C).

    The L/C contained the following provisions:

    a. Authorization to commence work immediately;

    b. A brief description of work;

    c. Prompt payments;

    d. $100K was initially authorized;

    e. The target date for contract finalization was 4 weeks after the execution of the L/C; and

    f. The L/C could be extended by mutual agreement.

2. After execution of the L/C (and commencement of work) the Firm prepared a draft contract (Contract). The Contract included the following:

    a. A detailed scope of work—jointly developed with the City's director of public works (DPW);

    b. Compensation and other non-standard clauses (e.g., invoice format, identification of key project personnel, requirement for a project office, responsibilities of the parties, etc.); and

    c. Standard contract clauses (i.e., "boilerplate"—e.g., liability, access to records, insurance requirements, etc.)

3. The Contract was submitted to the City's director of law (D/L) and a negotiation meeting was scheduled with the Mayor the following week. The president and a vice president represented the Firm at the meeting. The Mayor took exception to many of the Contract clauses (including the scope of work). The officers returned to their office and made the changes as requested by the Mayor.

4. At the second negotiation session (the following week), the Mayor took additional exceptions to the Contract (except for the scope of work). A third negotiation meeting was scheduled. In the meantime, the L/C had to be extended.

5. During the week (between the second and third negotiation meetings), the Firm's director of finance and administration (DF&A) and the D/L were in telephone contact. The Contract was reviewed and the "necessary" revisions made. After the third meeting, the officers returned with an unsigned contract and directions from the Mayor—to make more Contract changes. No additional meetings had been scheduled.

6. On Thursday morning (2 days after the third meeting), the Firm's president informs the DF&A of the following:

    a. The latest revisions the Mayor requested were being made.
    b. He has just talked to the D/L and the Contract "seems to be in order now."
    c. A meeting has been scheduled for tomorrow morning (9 a.m.) with the Mayor and the D/L. The purpose of the meeting is to get the Contract executed.
    d. The City will pay the invoices to date ($50K) and a check will be ready tomorrow.
    e. The DF&A should hand-carry the Contract to Knoxville, get the Contract signed, and pick up the check.

7. After spending Thursday night in Atlanta, the DF&A arrives in Knoxville at 7 a.m., has breakfast and promptly arrives at the City Hall for his 9 a.m. meeting. The DF&A meets the D/L and announces that he is there to get the Contract signed and pick up the check. The D/L says, "There must have been a misunderstanding. I told your president to send me the revised Contract for our review. If everything was in order, we would execute it and return a copy. We also would mail you a check next week. Furthermore, the Mayor is leaving within the hour for an appointment in Nashville and won't return until late this afternoon." (Remember, it's Friday.)

8. The D/L suggests that the DF&A leave the revised Contract with him and promises that he will give it his immediate attention so that this entire matter can be resolved early next week. According to the D/L, "There was really nothing else he could do at this time."

Assignment:
What would *you* do now?

Index

Closure, 59–64
  post-game niceties, 62–63
  timing, 59–61
  wrap–up, 64
Government (U.S.), 70–72
  areas (of negotiation), 71
  basics, 70
  process, 71
  recap, 71–72
International, 67–69
  information, 68–69
  process (step-by-step), 67–68
Issues, see Negotiation
Knowledge, 17–18
Location (playing field), 20–22
  advantages at your field, 21
  advantages away from home, 21
Maslow, Abraham, 26–27
Negotiation
  definition, 5
  government (U.S.), 70–72
  international, 67–69
  issues, 13–14, 39–40
  location, 20–22
  logistics, 22
  objectives, 9, 17, 31
  process, 5
  sport, 6
  team, 9–12, 17–18
  team rules, 12
  training, 19
Process
  government (U.S.), 71
  international, 67–68

Strategies, 9, 23–34
  assumptions, 24
  defensive playbook, 30–31
  information, 32
  leverage, 32–34
  needs, 26–28
  offensive playbook, 25
  perceptions, 24–25, 28
  problem solving, 28–30
  relationships, 23, 27
  time, 31–32
  trust, 28
Tactics, 9, 37–43
  agenda, 38
  concessions, 39–41
  credibility and believability, 38–39
  game plan, 37–38
  introductions, 37–38
  issues, 39–41
  resistance, 41–42
  scorekeeping, 42–43
  transition game, 43
  ultimatum, 41
Team, see Negotiation
Techniques, 44–55
  communications, 44–48
  listening to, 48–49
  play-acting, 51–55
  talking to, 45–48
  uncooperative opponent, 51
  watching (nonverbal
    communications), 49–50